Signals and Communication Technology

More information about this series at http://www.springer.com/series/4748

Srikanta Patnaik · Xiaolong Li
Yeon-Mo Yang
Editors

Recent Development in Wireless Sensor and Ad-hoc Networks

Springer

Editors
Srikanta Patnaik
Department of Computer Science and Engineering
SOA University
Bhubaneswar, Odisha
India

Yeon-Mo Yang
School of Electronic Engineering
Kumoh National Institute of Technology
Gumi
Republic of South Korea

Xiaolong Li
Electronics and Computer Engineering Technology
Indiana State University
Indiana, IN
USA

ISSN 1860-4862 ISSN 1860-4870 (electronic)
ISBN 978-81-322-2128-9 ISBN 978-81-322-2129-6 (eBook)
DOI 10.1007/978-81-322-2129-6

Library of Congress Control Number: 2014952781

Springer New Delhi Heidelberg New York Dordrecht London

Printed on acid-free paper

Springer is part of Springer Science+Business Media (www.springer.com)

Preface

A WSN may be described as a network of nodes that cooperatively sense and control the environment enabling interaction between persons or computers and the surrounding environment. Recent developments in networking and material science and nanotechnologies are the driving force for the overall development of large-scale wireless sensor networks (WSNs). In addition, these technologies have merged together to enable a new generation of WSNs that differ significantly from traditional wireless networks, which was implemented 5–10 years ago. Like any other advanced technologies, the origin of WSNs can be traced back to military applications. The first wireless network, which has a close resemblance to a recently used WSN, is the Sound Surveillance System (SOSUS) developed by the United States Military in the 1950s. This network used submerged acoustic sensors hydrophones, distributed in the Atlantic and Pacific oceans. The same sensing technology is still existing today and serving for the peaceful applications. Afterward during 1980s, the United States Defense Advanced Research Projects Agency (DARPA) started the Distributed Sensor Network (DSN) program to formally explore the challenges in implementing distributed/wireless sensor networks. Later on, scientific research communities as well as academia join hands to develop the WSN technology. Subsequently, government and universities began using WSNs for various applications, such as air quality monitoring, forest fire detection, natural disaster prevention, weather stations and structural monitoring, power distribution, waste-water treatment, and specialized factory automation, which were basically heavy industrial applications.

Present day state-of-the-art WSN has less deployment and maintenance costs, more rugged, and last longer, and they are now used for various applications at our homes, work places, bringing new sources of information, control, and convenience to our personal and professional lives. Efficient design and implementation of wireless sensor networks has become a hot area of research, due to the vast potential of sensor networks to enable applications that connect the physical world to the virtual world. This volume covers the recent developments in the area of Wireless Sensor and Ad-hoc Network. Potential applications for such large-scale

WSN exist in a various domains, such as health monitoring, home security and surveillance, and personal environmental monitoring, such as temperature and humidity.

In future, micro-fabrication technology shall bring down the cost of sensor nodes resulting in the pervasive use of wireless sensor networks with a large number of nodes. For the smooth deployment of the future WSN, researchers and designers are now engaged in solving the complex trade-offs among many application variables including deployment costs, hardware and software, system reliability, security, and performance. Wireless embedded system designers must also consider these trade-offs and make alternative decisions, such as transducer and battery technology choices, frequency of wireless operation, output power and networking protocols. The complexity of WSN design not only represents one of the most significant barriers to the widespread adoption of WSNs, but also provides an opportunity for hardware and software technology suppliers to add value. Another trade-off is also use of well established, standardized mix of hardware/software solutions for different WSN applications.

Srikanta Patnaik
Xiaolong Li
Yeon-Mo Yang

Contents

About the Editors

Prof. Srikanta Patnaik is presently serving as a Professor of Computer Science and Engineering, SOA University, Bhubaneswar, India. He holds Doctor of Philosophy in Engineering from Jadavpur University, India. He has published more than 80 research papers and articles in international journals and magazines of repute. He has supervised 10 research scholars for their Ph.D. degrees. He has completed various funded projects as Principal Investigator from various funding agencies of India. Presently, he is serving as an Editor-in-Chief of two international journals namely, *International Journal of Information and Communication Technology* and *International Journal of Computational Vision and Robotics*, published from Inderscience Publishing House, England and also an Editor-in-Chief of Springer Book Series on Modeling and Optimization in Science and Technology (MOST).

Dr. Xiaolong Li is working as an Associate Professor, Department of Electronics and Computer Engineering Technology, Indiana State University since 2008. He has done his Ph.D. in Electrical and Computer Engineering in 2006 from University of Cincinnati. He did his M.S. in Electronics and Information Engineering from Huazhong University of Science and Technology, China in 2002 and did his B.A. in Electronics and Information Engineering, from the same University of Science and Technology, China in 1999. His major research includes Wireless and Mobile Networks, Wireless Ad Hoc Networks, Wireless Internet, Modeling and Performance Analysis, QoS in Communication Networks, Wireless Mesh Network, Microcontroller-based Applications, Security in Wireless Networks. He has more than 30 International publications in International Journals and conference proceedings of repute. He has also contributed a Book Chapter entitled "Impact of Mobility on the Performance of Mobile Ad Hoc Networks and Performance Analysis of Mobile and Ad Hoc Networks", published by NOVA Publishers. He has association with the Institute of Electrical and Electronic Engineering (IEEE), American Society of Engineering Education (ASEE), Association of Technology, Management, and Applied Engineering (ATMAE) and Epsilon Pi Tau (EPT).

Prof. Yeon-Mo Yang is a Professor in the Department of Electronic Engineering, Kumoh National Institute of Technology, Gumi, Korea. He received the B. Engg. in the field of Electrical and Electronic Engineering from Korea Advanced Institute of Science and Technology (KAIST), Daejeon, Korea in 1990 and Ph.D. from Gwangju Institute of Science and Technology (GIST), Gwangju, Korea in 2006, in the field of optical networks. From 2005 to 2006, he was a post-doctoral fellow in North Carolina State University (NCSU), Raleigh in Networks Engineering and a short-time visiting scholar at UC Berkeley, Berkeley, CA, in Wireless Sensor Networks. From 2006 to 2008, he was in Daegu-Gyeongbuk Institute of Science and Technology (DGIST) as a senior researcher where he has been a project leader of the developments of wireless sensor networks. His current research interests include wireless sensor networks analysis and implementation both at the physical layer and at the networks layer; dynamic bandwidth allocation schemes in passive optical networks; and realizations of hybrid embedded systems connected to Under Acoustic Sensor Networks, Mechatronics and Information Technology.

Introduction

"Recent Developments in Wireless Sensors and Ad-hoc Networks" is an edited volume in the broad area of WSNs. It covers various chapters like Multi-Channel Wireless Sensor Networks, its Coverage, Connectivity, as well as Deployment. It also covers comparison of performance of various communication protocols and algorithms, such as MANNET, ODMRP, and ADMR Protocols for Ad hoc Multicasting, Location-Based Coordinated Routing Protocol and other Token-based group local mutual exclusion Algorithms.

Chapter 1 entitled "Multi-channel Wireless Sensor Networks" contributed by Amalya Mihnea and Mihaela Cardei, discussed issues and challenges related to multi-channel and multi-radio networks. They have classified the channel assignment schemes into static, semi-dynamic, and dynamic, and also discuss methods proposed in each category. They have presented other related issues such as primary users, network capacity, interference, topology control, and power and traffic aware protocols. They have explained the concept of multi-channel algorithms lucidly for designing additional algorithms for wireless sensor networks.

In Chap. 2 "Coverage, Connectivity and Deployment in Wireless Sensor Networks", Yun Wang et al. have introduced three fundamental problems, i.e., sensing coverage, network connectivity, and sensor placement/deployment in a wireless sensor network (WSN). They have covered the open problems in this area, which includes sensing coverage and connectivity analysis in three-dimensional WSNs, nonuniformly distributed WSNs, and mobile WSNs.

In Chap. 3 "Development of Home Automation System by using ZigbeX and Atmega128 for Wireless Sensor Networks", Nik Khadijah Nik Aznan and Yeon-Mo Yang presented a framework and a test-bed of Home Automation systems by implementing the cost-effective ZigbeX and Atmega128 with TinyOS. They have proposed a house model, which is able to control the lights and curtain depending on the light intensity measured by the photodiode on the ZigbeX.

In Chap. 4 "Efficient Coordination and Routing Protocol for Wireless Sensor and Actor Networks", Biswa Mohan Acharya and S.V. Rao have discussed about the problem of communication and coordination of various sensor nodes and proposed an efficient model based on geometric structure called Voronoi diagram. They have proposed a new protocol, which is based on clustering (virtual grid) and Voronoi region concept and they have given the simulation results which they claim outperforms in terms of throughput, packet delivery ratio, average delay, and normalized routing overhead.

Chapter 5 entitled "Performance Comparison of BEMRP, MZRP, MCEDAR, ODMRP, DCMP and FGMP to Achieve Group Communication in MANET" by M. Rajeswari et al. presents a comparative performance of six multicast protocols for Mobile Ad hoc Networks—BEMRP, MZRP, MCEDAR, ODMRP, DCMP & FGMP focusing on the effects of changes such as the increasing number of receivers or sources and increasing the number of nodes.

Chapter 6 entitled "Token based Group Local Mutual Exclusion Algorithm in MANETs" by Ashish Khanna et al. proposed a generalization of the group mutual exclusion problem based on the concept of neighborhood, which is named as group local mutual exclusion (GLME). They have also proposed a token-based solution of the group local mutual exclusion. The authors have claimed that their proposed method is the first token-based algorithm to solve group local mutual exclusion problem in MANETs.

In Chap. 7 "A Dual-band Z-shape Stepped Dielectric Resonator Antenna for Millimeter-wave Applications", Ashok Babu Chatla et al. have presented a dual-band z-shape stepped dielectric resonator antenna (DRA) for millimeter wave applications. The authors claimed that their design can be used for inter-satellite service applications, which operate at 65–66 GHz.

In Chap. 8 "OCDMA: Study and Future Aspects", Shilpa Jindal and Neena Gupta, discussed the future trend of OCDMA technique that highlighted on the newly developed three dimensional codes based on optical orthogonal codes and codes from algebra theory and their performance is evaluated on two models Model A and Model B.

In Chap. 9 "Focused Crawling: An Approach for URL Queue Optimization Using Link Score" by Sunita Rawat has presented a case of scaling challenges for traditional crawlers and search engines due to the expansion of the worldwide web and also proposed a method of efficient and focused crawling to enhance the quality of web navigation.

In Chap. 10 "An Optimized Structure Filtered-x Least Mean Square Algorithm for Acoustic Noise Suppression in Wireless Networks", Asutosh Kar and Mahesh Chandra have proposed an improved pseudo-fractional tap-length selection algorithm in context with the FX-LMS algorithm to find out the optimum structure of the acoustic noise canceller, which best balances the complexity and steady state performance.

Last but not least, in Chap. 11 entitled "An Exhaustive Comparison of ODMRP and ADMR Protocols for Ad hoc Multicasting" myself along with my colleague Ajit Nayak have presented a comparative study of two well-known protocols for wireless multicasting. One of the considered protocols is *On Demand Multicast Routing Protocol* (ODMRP) and the other one is *Adaptive Demand driven Multicast Routing Protocol* (ADMR). ODMRP is a mesh based protocol, whereas ADMR uses a tree-based technology for routing.

Chapter 1
Multi-channel Wireless Sensor Networks

Amalya Mihnea and Mihaela Cardei

Abstract In this chapter, we discuss some issues and challenges related to multi-channel and multi-radio networks. We classify channel assignment (CA) schemes into static, semi-dynamic, and dynamic and discuss methods proposed in each category. Other aspects presented are related to primary users (PUs), network capacity, interference, topology control, and power- and traffic-aware protocols. For a better understanding, some basic concepts related to wireless communication are explained. An understanding of multi-channel algorithms in general could help in designing additional algorithms for wireless sensor networks (WSNs).

Keywords Multi-channel · Multi-radio · Wireless sensor networks · Channel assignment · Primary user · Interference · Energy efficiency

1.1 Introduction

Wireless sensor networks (WSNs) constitute the foundation of a broad range of applications related to national security, surveillance, military, health care, and environmental monitoring.

Many results and channel assignment (CA) schemes proposed for wireless ad hoc networks and mesh networks cannot be directly applied to sensor networks, which have different characteristics such as smaller packet size, less powerful radios, or fewer radios. There are also differences related to energy source, power, computational capacity, and memory. The main type of communication used by

A. Mihnea (✉) · M. Cardei
Department of Computer and Electrical Engineering and Computer Science, Florida Atlantic University, Boca Raton, FL 33431, USA
e-mail: amihnea@fau.edu

M. Cardei
e-mail: mcardei@fau.edu

S. Patnaik et al. (eds.), *Recent Development in Wireless Sensor and Ad-hoc Networks*, Signals and Communication Technology, DOI 10.1007/978-81-322-2129-6_1

WSNs for data gathering is *converge cast* where data travel from many nodes (e.g., sensor nodes) to a single node called sink or base station (BS).

With a single-radio and a single-channel, WSNs cannot provide reliable and timely communication in case of high data rate requirements because of radio collisions and limited bandwidth. Therefore, designing multi-channel-based communication protocols is essential for improving the network throughput and providing quality communication services.

Multi-channel protocols consist of two major components: CA and medium access control (MAC). In some protocols, these two are combined: Channels are selected at every access to the medium. A good CA is one that reduces interference among concurrent transmissions, maximizes the capacity of the network, mitigates packet congestion within a single channel and in the case of primary users, and preserves robustness to the presence of a primary user (PU).

1.2 Challenges and Classifications of Multi-channel Protocols

MAC methods are designed for coordinating communication: sharing the wireless medium and alleviating conflicts. They are contention-based such as carrier-sense multiple access (CSMA) or schedule-based such as time division multiple access (TDMA), frequency division multiple access (FDMA), and code division multiple access (CDMA). Some other approaches use power control and directional antennas to further reduce interference [1].

CA schemes can be static, semi-dynamic, or dynamic. The static ones assign channels for permanent use, at deployment time or during runtime. Even if the assignments can be renewed, radios do not change their frequencies during communication. Semi-dynamic schemes assign constant channels to radios, but these can change during runtime while in the dynamic approaches, there is no initial assignment of channels to radios and the channels can change between successive data transmissions.

Some dynamic schemes use a dedicated control channel which can only be used for exchanging control messages (negotiations of the channels for data transmission) while data are exchanged using data channels. These approaches do not need time synchronization and are easier to implement.

Another classification can be related to implementation/execution. If the CA is done by a central scheduler, then the implementation is centralized. Otherwise, if nodes negotiate and more nodes are involved in assigning channels, then the implementation is distributed. The communication between devices involved in the distributed protocols is done by exchanging messages. Centralized approaches have limitations such as lack of a global common control channel to support centralized control and poor scalability due to the difficulty of capturing consistent global information in a dynamic environment [2].

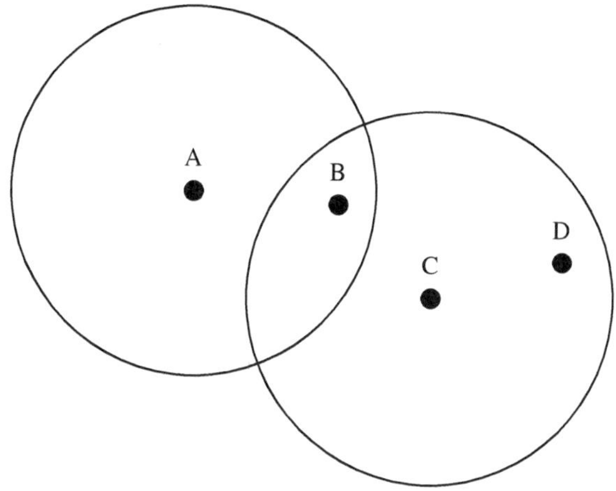

Fig. 1.1 The multi-channel hidden terminal problem

If switching channels is allowed, some problems that reduce network performance are as follows: multi-channel hidden terminal problem, deafness problem, and broadcast support [3].

The multi-channel hidden terminal problem (Fig. 1.1) can occur in carrier-sense multiple access with collision avoidance (CSMA/CA)-based protocols when the control packets (RTS/CTS) sent on a specific channel are not received by the nodes communicating on other channels. For example, suppose that we have four nodes A, B, C, and D that use a common control channel, let us say channel 1. Assume that A and B have successfully established a communication on channel 2. Assume that the node C was busy receiving on another channel when B sent the CTS to A, and thus, C is not aware of A communicating with B on channel 2. If subsequently C initiates a communication with D on channel 2, this will cause collisions at the node B. The cause of this problem is that nodes may listen to different channels when other nodes exchange RTS/CTS control messages on the common control channel [4].

In order to communicate, a sender and a receiver have to be on the same channel; otherwise, the deafness problem occurs in multi-channel communication. Suppose that a transmitter sends a control packet to initiate communication and that the receiver is tuned to another channel. If the sender does not get any response after sending multiple requests, then it may conclude that the receiver is not reachable anymore [3].

Broadcast support refers to the difficulty of supporting successful broadcasts when the nodes change their channels frequently. Usually, protocols use a broadcast channel to support broadcasts. If nodes operate on multiple channels, channel switching introduces delays and computational overhead.

Some additional problems that are mentioned in the literature are idle listening, when nobody is sending, and overhearing of messages, when a node receives messages that are destined to other nodes. It is known that the energy spent for receiving a message can even be a bit higher than the energy spent for transmitting a message [1]. All these problems are connected to power consumption, and they could shorten the lifetime of the network. They have to be taken into account when designing power-aware algorithms. Later, we will present some power- and traffic-aware protocols.

In [3] CA methods are classified using criteria such as assignment method, control channel, implementation/execution, synchronization, medium access, broadcast support, channel model, interference model, and objective. We elaborate on the classification into fixed (static), semi-dynamic, and dynamic methods, and we give a few examples for each category. The algorithms are assumed to be single radio unless otherwise specified.

1.2.1 Fixed Channel Assignment

An idea related to this category is to take advantage of *clustering* the nodes such that all the nodes in a cluster use a unique frequency that is different than the frequencies of other clusters. The goal is to avoid or to minimize interference.

Another idea is to use *component-based channel assignment* [5], which assigns the same channel to all nodes belonging to a component formed by nodes belonging to mutually intersecting flows. Two flows are said to be intersecting, if there is a common node in the set of active nodes for each flow, which serves both flows. If flow f_1 intersects with flow f_2 and flow f_2 intersects with flow f_3, then all nodes on the paths traversed by these three flows are assigned the same channel.

Different connected components can potentially operate on different channels. A connected component in a flow graph is defined as the largest subgraph, such that there exists a path from any node in the subgraph to all other nodes in the subgraph. The authors of [5] show through theoretical and quantitative analysis that this simple strategy can improve the network performance. They also propose centralized and distributed routing layer algorithms that implement this strategy effectively.

The main advantage of fixed CA is its simplicity since nodes maintain their assignments. But there are also disadvantages, such as not being adaptive to changes in the network topology due to traffic changes or unstable links, and no possibility of communication between two nodes that have different channels. These issues, which could lead to poor performance and network partitions, could be solved by renewing channel assignments from time to time.

The multi-hop scenario used in WSNs assumes that data travel from source nodes through intermediate nodes toward one or several BSs. Therefore, the routing topology is a tree or a forest.

In [6], a tree-based multi-channel protocol (TMCP) for data collection applications in WSNs is proposed. The authors assume that there is a single BS equipped with multiple radio transceivers, each of which works on different channels. The network is partitioned into multiple vertex-disjoint subtrees all rooted at the BS. Each tree is allocated a different channel, and a data flow is forwarded only along its corresponding subtree. TMCP is distributed and it works with a small number of channels, without any time synchronization requirement.

TMCP has three components: channel detection (CD), channel assignment (CA), and data communication (DC). Given *k* orthogonal channels, the CA module partitions the whole network into *k* subtrees, and one unique channel is assigned to each subtree. There is no inter-tree interference so the goal of partitioning is to divide the network into subtrees with low intra-tree interference. The DC component manages the data collection through each subtree.

TMCP uses the *protocol model* (a graph-based interference model) to estimate interference: Two nodes interfere with each other if the distance between them is smaller than a threshold value. The size of a node's interference set is used in subsequent subtree partition. However, the distance-based interference model does not hold in practice as shown by recent empirical studies.

In [7], TMCP has been extended to employ interchannel received signal strength (RSS) models for interference assessment in channel allocation. A novel algorithm is proposed, which can significantly reduce the overhead of multi-channel interference measurement by exploiting the spectral power density (SPD) of the transmitter.

In [2], three algorithms are presented: node-based, link-based, and node-link-based. The best of them, the node-link-based distributed algorithm, partitions the network into "stars," which resemble 2-level trees, and uses maximal matching between channels and adjacent links by the Hungarian algorithm. Channels are assigned to links that minimize channel-conflict probability by computing channel weights based on the conflict probability of every available channel on each link.

In dense networks, this algorithm could be used for the backbone consisting of cluster heads with longer transmission range while the communication within clusters could be done using CSMA. Cluster heads could solve inter-cluster interference by assigning inner-cluster communicating channels.

In [8], network robustness and channel interference are jointly considered when developing centralized and distributed algorithms. Backup channels are used to avoid network partition, but the requirement to adjust channels for previously assigned links might be unsuitable for WSNs, which have limited resources. The proposed solutions outperform existing interference-aware approaches when primary users appear and achieve similar performance at other times. The algorithms from these last two papers are not specifically designed for WSNs.

1.2.2 Semi-dynamic Channel Assignment

Methods in this category, which appear to be the most popular ones, assign fixed channels either to senders or receivers, but the assignments can change during communication. Graph coloring algorithms are useful in such approaches.

Network partitioning could be eliminated using a coordinated channel switching between senders and receivers, which need to be on the same channel at the same time in order to communicate.

In [9], a distributed game-based channel assignment algorithm (GBCA) is proposed to solve the problem of multi-channel assignment in WSNs. Unlike previous static assignment protocols, this algorithm takes into account both the network topology information and the transmission routing information. Simulations show that GBCA achieves better network performance than MMSN in terms of delivery ratio, throughput, packet transfer delay, and energy consumption.

The MMSN protocol [10] is the first multi-frequency MAC protocol designed specifically for WSNs, and it consists of two aspects: frequency assignment and media access. Frequency assignment allows users to choose one of four available frequency assignment strategies to evenly assign different channels among two-hop neighbors. In media access design, potential conflicts are solved by accessing the shared physical frequencies in a distributed way. Both GBCA and MMSN are distributed approaches.

1.2.3 Dynamic Channel Assignment

In these approaches, mostly distributed, a channel selection takes place before every data transmission. The channel selection can be measurement-based or status-based. The first category is related to communicating parties measuring signal-to-interference noise ratio (SINR) and the second one to the status of the channels: idle (available) or busy.

When the traffic is light, many multi-channel MAC protocols for WSNs are less energy-efficient than single-channel MAC protocols. In contrast to these, Y-MAC [11] is energy-efficient and maintains high performance under high traffic conditions. It is also the first protocol that uses dynamic channel assignment in WSNs.

In this TDMA-based multi-channel MAC protocol, a send time slot is used for data transmission and a receive time slot for data reception. An exclusive send time slot in two-hop neighborhood guarantees collision-free access to the medium, which reduces the energy wasted by contention and collisions. However, energy is wasted due to overhearing and idle listening, since all nodes have to wake at every time slot to avoid missing incoming messages. Initially, a base channel is used to exchange messages. Sensor nodes hop to the next radio channel if they have additional pending messages for the receiver (bursty traffic). Y-MAC improves the performance of the network (increased throughput, reduced message delivery latency) under high traffic conditions and uses multiple channels with low energy consumption. Other dynamic protocols such as MAC and MuChMAC are presented in [3].

1.3 Primary Users (PUs)

Due to the recent growth of wireless applications, the communication on the unlicensed spectrum (e.g., ISM) has become congested, while the utilization of the licensed spectrum varies between 15 and 85 % temporally and geographically [12]. Cognitive radio networks (CRNs) constitute a promising solution used to address the issue of inefficient spectrum usage.

A cognitive radio, also called a software defined radio because its communication functions are implemented on software instead of hardware, is a radio that can sense its environment, track changes, and react based upon its findings by efficiently avoiding interference.

Cognitive radios are designed to operate on a wide spectrum range and can switch to a different frequency band with limited delay. This technology allows PUs to share the spectrum with secondary users (SUs), where SUs communicate through un-assigned spectrum bands without disrupting the regular usage of the PUs. CRNs allow SUs to take advantage of unoccupied spectrum in an opportunistic manner using dynamic spectrum access strategies.

To avoid interference with a PU, a SU must vacate the spectrum when the channel is being used by a PU. This affects ongoing communication of the SUs. The challenge occurs due to the difficulty to predict when a PU will appear in a given spectrum. To use other channels, SUs have to spend a considerable amount of time for spectrum sensing and channel switching [13]. In addition, a change in a SU channel may trigger other nodes to change their channels in a ripple effect in order to maintain the desirable topology.

In the presence of PUs, the robustness constraint requires that if a channel is reclaimed by a PU, then the resulting SU topology still preserves the connectivity between any two nodes. The PUs can affect part of the network or the entire network (e.g., transmission of the TV tower). If two sensors u and v communicate on a channel that is reclaimed by a PU, then the packet is re-routed from u to v through another channel of u and possibly another radio of v. Thus, packet dropping and significant delays can be avoided.

1.4 Capacity, Interference, and Topology Control in Wireless Sensor Networks

A special type of WSNs are wireless multimedia sensor networks (WMSNs), which enable advanced surveillance, traffic monitoring, and healthcare systems, and require larger bandwidth. A challenge of WMSNs is an increased bandwidth demand in the presence of higher levels of interference. Using multiple channels for parallel transmissions could improve network capacity.

The most common communication types in WSNs are broadcast and data collection. The main goal of broadcasting is to send a message to all the nodes in the network, and the goal of data collection is to send data messages from source nodes to the sink(s). Broadcast could be used as an initial step in data collection to determine shortest paths between nodes and the sink(s). Communication in WSNs has to take several factors into account, such as link length, number of hops to the BS, and node degree [14].

For a given network topology, different routing trees or CA mechanisms have impact on the maximum achievable network throughput. With multiple channels, there is a need for channel coordination: The sender and the receiver have to transmit and to listen on the same channel at the same time.

In the *receiver-based channel allocation*, a fixed channel is assigned to each sensor node and that channel is used to receive messages. A neighbor that wants to send a message to this node should use the receiver's channel to send. In this allocation, the nodes that do not receive any message are not assigned any channel.

In the *link-based channel allocation*, every link or edge is assigned a channel and every transmission along that link uses that channel. A difference between this and the receiver-based channel allocation is that here, for the same receiver, various senders can use different channels, resulting in less interference.

In data gathering WSNs where source and sink nodes are all equipped with half duplex transceivers, the maximum throughput per node is W/n, where n is the number of source nodes and W is the transmission capacity. According to [1], the maximum throughput can be reached only if the sink is 100 % busy receiving packets and if the schedules of all nodes are aligned for interference-free communication for the given network topology.

In [15], it was proved that minimizing the schedule length for an arbitrary network in the presence of multiple frequencies is NP-hard. Also, finding the minimum number of frequencies that are necessary to remove all the interfering links in an arbitrary network is NP-hard.

The authors use an *aggregated convergecast model* [16] where each node has the ability to aggregate all the packets from its children as well as its own data into a single packet before transmitting it to its parent. The routing structure used in data collection is a tree rooted at the sink, and the frequency assignment strategy is receiver-based.

Each node has a single, half duplex transceiver, so it can either transmit or receive a single packet at any given time slot. The radio cannot receive multiple packets simultaneously so assigning different frequencies to the transmitters that are children of the same parent does not help in reducing the schedule length. A contention-free multiple access protocol such as TDMA is used, and each node generates only one packet at the beginning of every frame.

A graph-based interference model is used, where the interference range of a node equals its transmission range. Two types of interference for concurrent transmissions on two edges are considered: *primary interference*, if the two edges are adjacent, and *secondary interference*, if the receivers of both edges are on the same frequency and at least one of the receivers is within the communication range of the other transmitter.

The authors give an upper bound on the maximum number of frequencies required to remove all the secondary interfering links and also propose a polynomial time algorithm that minimizes the schedule length under this scenario. A secondary interfering link is removed if the two receivers of an edge pair are assigned different frequencies. Because half duplex radios are used, the primary interference cannot be removed using multiple frequencies.

A closely related work [16] describes a realistic simulation-based study on a tree-based data collection utilizing transmission power control, multiple frequencies, and efficient routing topologies. It was shown that the data collection rate becomes limited by the maximum degree of the tree once all the interfering links are removed by assigning multiple frequencies. This rate can be further increased on degree-constrained trees.

In [17], a multi-path scheduling algorithm for the snapshot data collection in single-radio multi-channel WSNs is proposed. A tighter lower bound for its achievable network capacity is given compared to the results in [18]. Also, a novel continuous data collection method for dual-radio multi-channel WSNs is shown to speed up the data collection process and improve the network capacity. Most of the previous works related to network capacity consider just single-radio single-channel WSNs. The protocol used in [18] is the *protocol interference model,* but the results can be extended to WSNs under the *physical interference model* [19].

The protocol interference model assumes that all nodes have the same interference range R. If a node X_i sends data to a node X_j over a channel, the transmission is successful if the destination node is far enough from the source of any other simultaneous transmission on the same channel or $|X_k - X_j| \geq (1 + \Delta)|X_i - X_j|$, where $\Delta > 0$, for any node X_k transmitting over that channel. This model can take advantage of the graph-coloring-based scheduling algorithms.

The physical interference model (SINR model) is considered better because it can capture the interference from multiple simultaneous senders. If $\{X_k; k \in T\}$ is the subset of nodes transmitting simultaneously at some point in time over a certain channel and P_k is the power level chosen by the node X_k for $k \in T$, then the transmission from a node X_i, $i \in T$, is successfully received by a node X_j if

$$\frac{\frac{P_i}{|X_i - X_j|^\alpha}}{N + \sum_{k \in T, k \neq i} \frac{P_k}{|X_k - X_j|^\alpha}} \geq \beta$$

where β is the minimum SINR for successful receptions, α represents the exponent for signal loss due to distance, and N is the level of the ambient noise [3].

In [20], the authors analyze the capacity limits of multi-hop paths in a WSN with multiple channels. Also, a control channel-based MAC protocol is implemented and analyzed using IEEE 802.15.4-based networks with 16 orthogonal channels. This protocol is based on the split-phase approach described later, which does not require any time synchronization and is simple to implement, but could suffer from saturation of the control channel.

Previous MAC protocols such as McMAC, CMAC, and MAC are not considered efficient. The channel coordination mechanisms are divided into four categories: (1) dedicated control channel, (2) common hopping, (3) split phase, and (4) McMAC. The differences of these mechanisms are related to the number of radios they use: a single radio or two radios.

The dedicated control channel mechanism is used with multiple radios. There is a specified control channel used for one radio to transmit information related to channel selection. After a channel is selected, data are transferred through that channel between the sender and the receiver.

The common hopping mechanism is used with a single radio. There is a common pattern of shifting channels followed by all nodes. An RTS/CTS handshake takes place using the current common channel when two nodes want to communicate. During this handshake, the given sender–receiver pair stops hopping and stays tuned to this common channel. After data transfer is done, the two nodes resume the same hopping sequence.

The split-phase approach is used with a single radio and has two phases: control and data. The control phase includes agreements that are made between sender(s) and receiver(s) regarding the channel on which data have to be transferred. In the data phase, the data are transmitted using the chosen channel.

The fourth category includes the McMAC [21] protocol, which uses random hopping of channels for each node. When a node A has to transfer data to a node B, if data to be transferred are large, then A follows B's hopping pattern. After data transfer, A resumes its own hopping sequence.

In addition to MAC protocols, other methods to alleviate interference are as follows: transmission power control (transmitting signals with sufficient power instead of maximum power) and use of directional antennas instead of omnidirectional antennas.

Topology control is a very important technique in WSNs which deals with sensor nodes' power control and network structure. Some of the design goals of topology control are as follows: minimum energy consumption, low interference, small node degree, connectivity, and planarity.

Some topology control algorithms that have been proposed in the last few years are inappropriate because they do not address both communication types in WSNs: message dissemination and data collection. Very often, the robustness of the topology is neglected. The node failure is handled by most of the algorithms by simply resetting the whole network, which has a high cost in terms of energy consumption. Other algorithms try to establish several disjoint routes between sensor nodes and the BS or between sensor nodes in order to improve the network robustness, which is not an easy task.

The simplest topology control strategy is called unit disk graph (UDG), in which all sensor nodes communicate with each other using their maximum power so that all possible communication links are preserved. Other approaches try to eliminate all or some of the redundant links, keeping some links to improve the network's tolerance to node failure and network capacity, or try to improve the robustness of the network by a specially designed edge weight function, which contains link length and number of hops to the BS.

Other topology control algorithms partition the network into several disjoint parts or clusters, each of which has one selected cluster head and several cluster members. The cluster head is responsible for inter-cluster communication, and all cluster members only communicate with their own cluster head in a TDMA manner. Each cluster head is elected periodically in order to balance energy consumption.

In [14], the authors propose a novel tree-based topology control algorithm which contains a fast dissemination tree (FDT) for message broadcast and a balanced data collection tree (BDCT) for data collection. This algorithm has better performance than the existing ones and helps balance energy consumption between nodes.

FDT uses the maximum power and chooses the nearest neighbor for transmission. Node i chooses as its parent the nearest node j which has the smallest number of hops to the BS, so that the message can be received and relayed as quickly as possible. The BS is placed in the center of the network, and each node has information about its neighbors, such as ID, number of hops to the BS, energy left, and location, which are used in the construction of the BDCT.

The objective of the BDCT is to achieve a balance among different design goals: link length, number of hops to the BS, remaining energy, and robustness. The selection of a parent node is based on the link weight whose settings give priority to a node with more residual energy. This helps balance the energy consumption throughout the whole network, which in turn prolongs the lifetime of the network.

Robustness is related to the number of critical nodes. Node i is said to be a critical node if, once i fails, the network is no longer connected. A tree-based topology has a severe drawback: once a parent node fails, all its children lose connection with the BS. A solution to this case is topology reconstruction, but this approach is expensive in terms of energy and time. Another solution is to establish multiple paths between a node and the BS, but computing multiple paths increases the computational complexity. A better solution proposed in [14] is to choose a network topology that resembles a spider web which has to be reinitialized just in the case of multiple node failures.

1.5 Power- and Traffic-Aware Protocols

Many WSNs applications need data to be transmitted in a timely manner, such as WSN-based disaster warning systems or a warning surveillance system that has to notify authorities when intruders are detected. It is known that wireless links are lossy and retransmissions increase the end-to-end delay. A solution to improve link quality is to increase transmission power, but this may increase interferences and channel contention, which leads to a decreased network capacity. An alternative solution is to use multi-channel protocols to increase network capacity and to reduce interference and delays.

In some MAC protocols, the energy efficiency is improved by reducing idle listening. Nodes sleep and wake up when they receive a message. Other protocols use a distributed time slot selection mechanism for scheduling senders or receivers. In LMAC [22], each sender owns an exclusive time slot in a two-hop neighborhood, and all nodes wake up at every time slot to avoid missing incoming messages. Receivers are scheduled in Crankshaft [23] as follows. Each node wakes up for data reception at a different offset from the start of the super frame in order to reduce the number of nodes overhearing unrelated messages.

Energy consumed to receive a message is considered to be greater than that to transmit a packet due to the sophisticated de-spreading and error correction techniques. Therefore, scheduling receivers is more energy-efficient than scheduling senders if the traffic is light, because each node samples the medium only in its own receive time slot [11].

A multi-channel protocol based on LMAC is presented in [24]. First nodes communicate on a basic channel, and when all the time slots are exhausted, new channels are used. Two nodes that are on different channels communicate through bridge nodes, even if they are within one hop from each other. This can increase latency and energy consumption by bridge nodes.

In [25], the network is clustered, and each cluster head collects request messages from the cluster members and then assigns channels to both source and destination nodes. One disadvantage of this method is that cluster head nodes consume more energy than the other nodes. Also, the maximum network throughput of each cluster is limited by the number of packets that the cluster head can manage.

Some MAC protocols that focus on the efficient use of energy do not address bursty traffic, which became a major issue due to the ability of the latest operating systems for WSNs to run multiple applications. Newer MAC protocols combine the advantages of contention-based protocols and time-slotted protocols.

The multi-radio MAC protocol presented in [26], which uses two radio transceivers on each sensor node, separates the transmission of control packets and data packets into low- and high-frequency bands, respectively. The information necessary for selecting high-frequency channels is exchanged in micro-preamble frames using the low-frequency band. Data transmission will be then carried out using the selected channel. The high-frequency band radio is turned on only for a very short time for burst data transmissions. The control channel is also used for small data packets. The multi-radio protocols are not considered an economical solution in terms of energy, and most approaches focus on devising a multi-channel MAC protocol using a single-radio transceiver.

In [27], a multi-channel real-time (MCRT) communication protocol that uses both multiple channels and transmission power adaptation for real-time WSNs is presented. MCRT uses a small number of orthogonal channels that are allocated to network partitions formed based on many-to-one data flows in a WSN. The goal is to minimize the channel contention among different flows.

This algorithm outperforms the one presented in [28], called the real-time power-aware routing (RPAR) protocol, which dynamically adapts its transmission power and routing decisions based on packet deadlines. Practical issues in WSNs, such as lossy links, scalability, and severe memory and bandwidth constraints, are also addressed by this algorithm. A novel neighborhood manager dynamically discovers eligible forwarding choices and manages the neighborhood table. New forwarding choices are discovered by adapting the transmission power to a neighbor that is already in the neighbor table (*power adaptation*) or by discovering new neighbors (*neighbor discovery*).

This paper handles smaller holes through power control and larger holes through face routing mechanisms. Holes may appear in the network topology due to voids in node deployment or node failures.

A protocol that deals with power management issues has to minimize the energy consumed for packet transmission and reduce the energy wasted on idle listening. The real-time power-aware routing (RPAR) protocol is discussed in connection with several power management techniques.

In [1], reinforcement learning is used to make sensor nodes learn how to achieve successful transmissions and receptions while minimizing energy consumption. This distributed algorithm reduces the energy wasted by collisions, idle listening, and deafness problem. The protocol is based on a combination of TDMA and FDMA techniques.

Time is divided into frames of fixed length, composed of a number of time slots. The length of a slot allows the transmission of a single data message and an acknowledgment message. A channel-hopping scheme is used instead of a fixed frequency assignment for each node.

In each frame, a node periodically switches its channel at each time slot according to a chosen channel-hopping pattern, called the *default sequence*, obtained with a pseudorandom number generator. When establishing communication, the sender tunes its radio to the current channel of the receiver and transmits data.

When it knows the receiver address, which is the seed of its generated default sequence, the sender can reproduce the listening channel of the receiver. With this method, parallel transmissions between several pairs of nodes can take place without exchanging information or negotiating a communication channel, which represent a major challenge for energy- and bandwidth-constrained WSNs.

Each node in the network has multiple parents that it can use to forward data to the sink and it decides which one to use based on which channel the parent is listening to. A node can choose from the following pool of actions: stay on its home channel and listen, deviate to a chosen channel to transmit in a certain time slot of the frame, or withdraw from communication in order to save energy. The main objective of the scheduling algorithm is to coordinate the action of each node in each time slot in order to reduce failed transmissions.

First, each node chooses a random action for each slot. The actions are performed simultaneously, and the nodes that were successful in establishing communication receive positive feedback. If an action was successful, it will be repeated in the same slot in the next frame (“win-stay”); otherwise, an action from the set of available actions will be randomly selected when the executed action failed (“lose-shift”).

The actions could be chosen using a *uniform* or *biased* probability distribution. With the uniform scheme, all actions in the set of available actions have equal selection probability. With the biased distribution, the probability of an action to be chosen as the next action to perform is exponentially proportional to the probability of successfully performing that action (this is updated in every frame).

To avoid interference with an already successfully coordinated communication, a clear channel assessment (CCA) check is done by nodes before transmitting. At the beginning of each slot, a contention window is used for performing CCA. The length of this contention window is decreased for nodes that have previously established successful transmissions in that slot. Nodes that performed successful transmissions will get higher priority with this mechanism.

This protocol is compared with McMAC [21] which is a frequency-hopping protocol with multiple rendezvous. The proposed collision-free schedule protocol performs better than McMAC since McMAC does not provide contention-free access. It is also more energy-efficient compared to McMAC.

In [1], the authors define α as the ratio of the number of time slots per frame to the number of generated data messages at all source nodes in the network. For large enough α, the optimal solution for their proposed contention-free multi-channel protocol is achieved more easily and more often, and also the latency and the packet loss are reduced. But increasing α will result in more idle listening. When the number of slots per frame is larger, nodes might try to transmit in more slots before the receiver node concludes that the action of listening is a good one.

The performance of the algorithm also depends on the sleep threshold, which should not be set too large, otherwise a lot of nodes might decide to sleep. The protocol uses a schedule-based approach and requires time synchronization for the whole network.

The scheme presented in [29] uses two metrics with two separate goals: to minimize the energy consumption and to minimize the delay when delivering packets to the destination. The proposed scheme can adapt to various types of traffic occurring in the network, and it tries to minimize both energy and delay for the respective traffic classes. Two types of packets are considered: urgent and normal packets where urgent packets need to be delivered with a minimum delay to the end user, while normal packets do not have this requirement and therefore are delivered with a minimum energy.

This scheme could be used for example in environmental monitoring applications, where sensed data that are not critical or urgent can be forwarded with a minimum energy, while those having more important information (such as a fire detection) must be delivered faster to the destination.

An important issue in WSNs is node failures. Path discovery is usually done by flooding. When more paths from a node to the sink are maintained, the network can usually recover from failures on the primary path without invoking network-wide flooding for path discovery. This is very important in sensor networks because flooding involves energy consumption, and this reduces network lifetime.

In [30], multipath routing in WSNs is studied using two different approaches to construct multi-paths between two nodes: one with disjoint paths, in which the alternate paths do not intersect the primary path or each other, and one with braided paths, in which there are no completely disjoint paths, but rather many partially disjoint alternate paths.

The *resilience* of a scheme refers to the likelihood that, when the shortest path fails, an alternate path is available between a source and the sink. But with multiple paths, there is *maintenance overhead*, which is considered a measure of the energy required to maintain these alternate paths using periodic keep-alives. Becoming more resilient typically consumes more energy so there is a tradeoff here.

Two different failure modes are considered in [30]: *isolated* node failures, where each node has an independent probability of failure, and *patterned* failures, in which all nodes within a certain fixed radius fail simultaneously. For the same patterned failure resilience, the braided multi-paths are preferred over disjoint paths because they have about 50 % higher resilience to isolated failures and a third of the overhead for alternate path maintenance. Also, it is considered that disjoint alternate paths are hard to construct in localized algorithms due to lack of information.

DRCS [31] is a multi-channel protocol that performs channel selection and routing together in order to improve the battery lifetime in WSNs. A node from the network has a receiver channel (the channel used for receiving all incoming packets) and a transmit channel (the channel to which a node temporarily switches to transmit a packet), which is the receiver channel of its intended destination. Nodes listen on their receiver channels by default.

Selection of channels and parents (nodes to send to) is done based on a battery health parameter H and a path metric that is calculated using a link quality parameter (ETX). The battery health metric H of a node represents its remaining battery lifetime which depends on its currently estimated energy usage. The quality of a route is estimated using a path metric that is calculated as the sum of the expected number of transmissions (ETX) on each of its links. An ETX for a link is defined as the expected number of transmission attempts required to deliver a packet successfully over the link.

A node chooses the route with the lowest path metric to the sink. When choosing the transmit channels, the goal is to prolong the lifetime of the neighboring node with the worst battery health metric. Channel selection is connected to the parent selection, and this will determine the route that the message takes to get to the destination; therefore, the proposed approach leads to a joint channel selection and routing in the WSNs. DRCS generates a higher packet delivery ratio in comparison to TMCP.

Traffic patterns change significantly during runtime, and some nodes or segments of the network may have more traffic than others. CA algorithms that do not take traffic patterns into account may waste channels on nodes with no traffic while assigning too few channels to nodes with heavy traffic. CA strategies that can dynamically adapt to traffic pattern changes are desirable.

The traffic-aware CA algorithm presented in [32] collects traffic information from two-hop neighbors and uses a specific algorithm to assign channels among two-hop neighbors. This traffic-aware frequency assignment is incorporated into the existing MMSN MAC, and it is compared with two conventional frequency assignment methods: even selection and eavesdropping.

An important aspect of this algorithm is assigning a traffic weight to each node, from which we could infer its future reception data rate. The goal is to minimize the maximum load of any channel within the two-hop neighborhood of any node in order to decrease the number of radio collisions. This problem is similar to the load balancing job scheduling problem, which is NP-hard.

This traffic-aware algorithm is a greedy algorithm. First, nodes exchange information so that each node knows its two-hop neighbors' IDs and traffic weights. Then, nodes choose their channel in the decreasing order of their traffic weight: The node with the greatest traffic weight among its two communication hops selects the available channel with the least load and then beacons the channel choice in its two-hop neighborhood. The corresponding channel is then updated. Before they make their decision, nodes wait for nodes with greater weight and for nodes with equal weight but lower node ID.

This algorithm uses just one radio and can only use one channel at one time. Two groups of experiments were considered, by varying the system loads and the number of available channels, respectively. Simulation results show that this algorithm greatly improves multi-channel MAC performance, especially when the number of channels is medium, such as 4, 6, 8, and when the system load is light or heavy. It significantly enhances the packet delivery ratio and throughput, while reducing channel access delay and energy consumption.

This traffic-aware CA algorithm has a more efficient channel usage than the algorithm in [33] which also has the capability of dynamically changing the radio frequency. In this algorithm, nodes make local decisions to dynamically change the radio frequency. A *home frequency* is assigned to each node such that the network throughput is maximized. Initially, all nodes have the same home frequency (channel), and when the channel becomes overloaded, the nodes move to another channel.

When a node wants to send data to a node that has a different home channel, it temporarily switches to that channel to send the data. The main rule of this algorithm is to "cluster" into the same channel the nodes that communicate frequently and to separate into different channels those that do not. The protocol is distributed, and a node decides to switch channels based on an estimated communication success probability that is obtained after nodes exchange state information.

The switching is done based on a probability, and two goals are alleviating congestion and avoiding having all nodes jump to the new channel. The nodes that behave predominantly as sinks have priority to switch channels first (i.e., initiate the cluster split). If a node communicates heavily with another node that already switched, then it follows the node into the new cluster, and in this way, the communication between different clusters is minimized. Unlike other algorithms, this algorithm does not neglect the time to switch between two channels.

A novel interference-aware multi-channel media access control (IMMAC) protocol which can be easily implemented in WSNs is presented in [34]. This single-interface multi-channel protocol addresses issues such as dynamic traffic, non-negligible channel switching overhead, and multi-channel utilization support for broadcast. IMMAC does not require time synchronization.

Assuming that the relative traffic on each link of the network is known a priori, each node is assigned an initial channel for data reception such that the maximum concurrency is achieved. Channels are assigned three priorities when choosing the initial channel for each node: high priority, middle priority, and low priority. Channels that are not used in the two-hop neighborhood have the high priority. Channels not used by two-hop away neighbors but used by one-hop neighbors have the middle priority. The other channels have the low priority. The initial channels are adjusted according to dynamic traffic.

Each node maintains a local channel assignment table (CAT), in which it records the node ID and channels assigned among its two-hop neighbors. To maintain fairness on every channel, the scheduler uses a round-robin method to transmit on each nonempty channel queue. A node maintains a nonempty queue channel to transmit until the queue becomes empty, or the staying time is longer than a threshold value $T_{\max}$, which depends on the tradeoff between switching overhead and packet delay. The protocol uses receiver-based channel switching to support unicast and broadcast communication.

The traffic is not always known prior to CA, and it could also change over time. CRNs are able to perceive the environment and adapt according to the conditions of the network (neighboring nodes' channels, traffic demand, etc.). A cognitive process is able to make future decisions based on the success of past decisions and to adapt its CA when traffic changes.

Next, we present some aspects related to radios and multi-radio networks. The previous algorithms were mainly single-radio multi-channel.

1.6 Radio and Multi-radio Issues

Battery life of sensor nodes is a very important issue in WSNs, and energy efficiency could be obtained by designing sensors that consume low power, by energy harvesting, and by efficient scheduling at the node and network level. Most of the energy is spent on wireless communication.

There are two types of data transmission in traditional wireless networks: unicasting (one-to-one) and multicasting (one-to-many). In addition to these, in WSNs, several data source nodes sense the data and send them back to the sink node. This is called *reverse multicasting* (many-to-one) or *data aggregation routing*.

If two children sensor nodes use two different channels to transmit data to the same sensor node, then this sensor node will need two radios to receive data simultaneously. Otherwise, a single radio will have to switch its channel to receive from its children nodes, and this will increase the latency. Therefore, for minimum latency, any sensor node that is on the data aggregation tree has to have a number of radios that is greater than or equal to the number of children nodes. Each radio is assigned to a channel from each sending node.

The same is true for sinks: A sink with a single radio that receives data from nodes on different channels has to constantly switch channels, so multiple radios are needed to improve the performance at the sink. It was shown that if we increase the number of sinks and/or increase the number of radio interfaces in the sink, then we can obtain a better performance at the sink, which results in an overall performance improvement within the network [35].

There are different types of radios in a multi-radio sensor node that may differ in terms of their communication capabilities, energy efficiency, and usage. High-bandwidth, long-range radios are more energy-efficient in terms of energy expended per bit transmitted than low-bandwidth, short-range radios. But high-bandwidth radios consume more power than low-bandwidth radios when they are idle.

Therefore, when there is not a lot of data to be transmitted, it is not recommended to activate many high-bandwidth radios because this may waste energy. On the other hand, long-range radios have a greater reach and can reduce the network diameter. Using long-range radios will decrease the latency involved in delivering sensory data to a specific destination. Radios with higher power and longer range may need to be activated in order to achieve a connected network.

In networks with multiple types of radios, the problem of *energy-efficient radio activation* is to minimize the total energy spent by the active radios across all nodes in order to maintain a connected network. This problem is NP-hard, and four approximation algorithms are presented in [36].

Most of the work on multi-radio networks was not focused on WSNs, and the results cannot be directly applied to them. The approaches could be divided into two categories: centralized and distributed. Centralized approaches can further be divided into flow-based, graph-based, and partition-based.

When assigning channels, some approaches give priority to high load edges or to the edges that are closer to the sink. The routing algorithm uses shortest path routing or sets of paths between two communicating nodes. Most approaches are not realistic because they assume that the traffic is constant, while network traffic can be bursty.

Channel scheduling in multi-radio multi-channel (MR-MC) wireless networks has been studied under the assumption that the communication range equals the interference range. In reality, this assumption cannot be satisfied because a node's interference range is usually larger than its communication range. This means that two nodes might interfere with each other even if they cannot communicate directly with each other.

Under the *physical interference-free model*, two nodes interfere with each other if and only if their physical distance is less than or equal to the interference range. Under the *hop interference-free model*, two nodes interfere with each other if and only if their hop distance is less than or equal to a number of hops H. When node location is unavailable, the hop interference-free model can be applied, even if it is not as precise as the physical interference-free model.

In [37], the authors proved that channel scheduling is NP-hard under both of these models in MR-MC wireless networks. Also, they proposed a polynomial time approximation scheme (PTAS) framework that could be used for channel scheduling under both interference models in such networks.

The factors that affect the energy consumption of a radio are as follows: the sensor node hardware architecture, the background network traffic, and the network topology. The authors of [38] draw conclusions related to designing energy-efficient data transmission protocols that are congestion aware. Their experimental results show that always transmitting packets of maximum size can have a negative impact on radio's energy consumption. By properly adjusting the runtime parameters, such as the packet size and the packet generation period (packet rate), the radio's energy consumption can be significantly reduced under heavy network traffic.

Some approaches use two or more radios with different energy and throughput characteristics to minimize the total energy consumption. In [38], the authors studied a heterogeneous sensor node equipped with 802.15.4 and 802.11b radios and performed an extensive set of experiments, while varying various network parameters. They conclude that by properly adjusting some network parameters, such as packet size and transmission period, energy savings of up to 50 % can be achieved under heavy network traffic conditions when a CSMA-based MAC is used. Also, a proper pairing of processor and radio is crucial for taking full advantage of the energy efficiency of higher bandwidth radios.

Chipcon's 802.15.4 compliant radio is a low-power radio that can provide data rates of 250 Kbps, which are sufficient for many sensor network applications. High-bandwidth radios such as 802.11b have higher power consumption, but they provide significantly higher data rates (11 Mbps). This means that high-bandwidth radios can transmit more data in less amount of time compared to low-power radios, and they can be more energy-efficient. For example, the energy per bit of the 802.15.4 radio (979 nJ/bit) is almost 9 times higher than the energy per bit of an 802.11b radio (112 nJ/bit) [39].

In addition to the energy per bit metric, there are other system aspects that affect radio's energy dissipation. Besides high power consumption, radios such as 802.11b have a large start-up time. The time it takes to power up and configure the radio is orders of magnitude higher (approximately 2–3) compared to that of low-power radios such as 802.15.4 (less than 2 ms). So every time we power up the radio, a fixed energy overhead is created, which is independent of the size of data to be transmitted.

Taking into account this high start-up cost, high-bandwidth radios are more energy-efficient only when a large number of bytes have to be transmitted because the high start-up cost gets amortized as more and more bytes are transmitted. These radios should not be used if there is little or no data to send or if data need to be sent only occasionally.

The low-bandwidth radios are less energy-efficient but consume much less energy when idle. Also, they are able to quickly change their state from sleep to active in order to send data and then deactivate. Therefore, they are suitable for transmitting small amounts of data as well as remaining "vigilant" for long time periods.

In [39], the dual-radio system has the following characteristics: The primary processor and high-bandwidth radio remain off until triggered by the application, while the secondary processor and radio operate on low power and remain vigilant. The secondary radios form a network that is always available, while the primary network exists only when it is needed. We assume that the low-bandwidth network is connected and not partitioned, but the two network topologies are not necessarily the same. A neighboring node over the low-bandwidth radio is not necessarily a neighboring node over the high-bandwidth radio and vice versa.

In the centralized approach, there is an always-on node called *topology controller* (possibly colocated with the sink) that handles requests for end-to-end paths received from the low-bandwidth radios. Based on stored information about pre-established routing paths, this node sends wake-up requests to low-bandwidth radios, which turn on their CPU and high-bandwidth radio for data transfer. Just the nodes that are required for the path are woken up. In the distributed approach, each node keeps in non-volatile memory a path to all potential destinations in its routing table.

If a node fails, all the paths that contain that node are invalidated. When the controller wants to send a wake-up message to a node, if no path is found, the wake-up request is sent to all the nodes in the network.

In both approaches, the low-bandwidth radios communicate using routing trees or flooding. An important advantage of the centralized approach in terms of energy is the fact that if multiple concurrent transfers are involved, then the controller does not have to turn on any extra nodes if the network is sufficiently connected. This algorithm provides energy savings of more than 60 % compared to alternative approaches while incurring only a moderate increase in application latency.

The conclusion of [40] is that the pairing of two complementary radios with different range characteristics (maximum ranges 80 m vs. 800 m) enables greater range diversity at lower energy cost than a single radio. The radios are divided into two main categories: a long-range radio that enables communication over long distances but requires more power and a shorter range radio that is more power-efficient but cannot provide communication over longer distances.

Experimental results show that choosing a long-range radio and using its lower power settings for short-range communication is far less efficient than using a short-range radio. Directional antennas can increase the communication range of short-range radios, but directional antennas are bulky, and they are not suitable for many mobile WSNs.

The new multi-radio sensor platform used in [40] is called the arthropod, and it pairs two radios—CC2420 and XE1205—that offer diversity in frequency, power, and range, and it is a good solution for range-adaptive mobile WSNs. Also, a unified link layer is presented that, with the help of an adaptive algorithm based on reinforcement learning, chooses which radio to use for communication between a pair of nodes. The reinforcement technique that is used is Q-Learning, which provides a simple reward for making correct decisions and an ability to explore other operating points periodically [40].

This algorithm introduces energy and resource overhead as it has to continually monitor and learn channel characteristics for the two radios and determine which one is more efficient in terms of energy. Its main advantages are as follows: up to 52 % improvements in energy efficiency over using only one of the two radios on the platform and the ability to easily implement the learning algorithm with limited memory and computational overhead on a mote-class sensor platform.

A two-stage multi-radio algorithm that could adapt to traffic demand is presented in [41]. Using a greater number of radios increases the cost in terms of energy consumption, so this algorithm initially assigns a minimum number of transceivers to provide network connectivity over multiple channels (traffic independent or TI channel assignment) and then enables additional links to respond to increased traffic demands (traffic dependent or TD channel assignment).

Some literature discusses the optimal placement of nodes and the challenges resulted from the fact that obstructions to the line-of-sight between nodes are usually dynamic, so the optimal placement cannot be decided at the deployment time. Also, connectivity at deployment time can change over the network lifetime. WSNs have to adapt to changes in network conditions, and the existence of alternative routes is helpful. CA has to take these aspects into consideration.

IEEE 802.15.4 standard uses fixed CA, but the beacon node can change the operating frequency of its network if the nodes report excessive levels of interference. There is also literature that explores interference between coexisting networks. A solution for avoiding interference is adjusting the transmission power to a level that is sufficient to reach the receiving node.

1.7 Conclusions

Using multiple channels and multiple radios in WSNs, we can achieve a better support for applications which require high network throughput.

To provide efficient communication, sinks and cluster heads need multiple radios so they can communicate with multiple nodes simultaneously. Also, they need sufficient power supply. Nodes in a cluster could take turns in assuming the role of cluster head.

A disadvantage of the protocols that use a control channel is the existence of a potential single point of failure: the control or primary interaction channel. The connectivity is lost if there is interference or severe fading on this channel. Also, in the presence of PUs that reclaim specific channels, the connectivity of the network has to be maintained.

The distributed protocols are considered more practical and suitable for real-world applications than the centralized ones. In the case of WSNs, it is important to balance the energy consumption, and this is achieved by distributing the work load among sensors with the purpose of equalizing their remaining lifetimes. Some traffic could be redirected so as to ensure a balanced utilization of all the sensor nodes.

The network has to adapt to failure of nodes which could deactivate paths between a node and the BS and determine changes in traffic. Maintaining multiple paths is useful in such cases, but this increases energy consumption. In the presence of PUs, the availability of channels changes, so CA algorithms have to adapt accordingly.

The use of multiple channels in WSNs allows parallel transmissions that improve the performance of the network, resulting in increased throughput, decreased collision probability, and thus better energy efficiency. Because WSNs are limited in terms of power and processing capabilities, developing simple algorithms that could be scaled to large networks is of major importance in improving the performance of such applications.

References

1. Phung, K.-H., Lemmens, B., Mihaylov, M., Tran, L., Steenhaut, K.: Adaptive learning based scheduling in multichannel protocol for energy-efficient data-gathering wireless sensor networks. Int. J. Distrib. Sens. Netw. (2013)
2. Wu, J., Dai, Y., Zhao, Y.: Effective channel assignments in cognitive radio networks. Comput. Commun. **36**(4), 411–420 (2013)
3. Incel, O.D.: A survey on multi-channel communication in wireless sensor networks. Comput. Netw. **55**(13), 3081–3099 (2011)
4. So, J., Vaidya, N.H.: Multi-channel MAC for ad hoc networks: handling multi-channel hidden terminals using a single transceiver, In: MobiHoc '04: Proceedings of the 5th ACM International Symposium on Mobile Ad Hoc Networking and Computing, Roppongi Hills, Tokyo, 2004
5. Vedantham, R., Kakumanu, S. Lakshmanan, S., Sivakumar, R.: Component based channel assignment in single radio, multi-channel ad hoc networks. In: MobiCom '06: Proceedings of the 12th Annual International Conference on Mobile Computing and Networking, Los Angeles, CA, 2006
6. Wu, Y.F., Stankovic, J.A., He, T., Lin, S.: Realistic and efficient multi-channel communications in wireless sensor networks. In: INFOCOM '08: Proceedings of the 27th IEEE International Conference on Computer Communications, Phoenix, AZ, 2008
7. Xing, G., Sha, M., Huang, J., Zhou, G., Wang, X., Liu, S.: Multi-channel interference measurement and modeling in low-power wireless networks. In: RTSS '09: Proceedings of the 30th IEEE Real-Time Systems Symposium, Washington, DC, 2009
8. Zhao, J., Cao, G.: Robust topology control in multi-hop cognitive radio networks. In: INFOCOM '12: Proceedings of the 31st IEEE International Conference on Computer Communications, Orlando, FL, 2012
9. Yu, Q., Chen, J., Fan, Y., Shen, X., Sun, Y.: Multi-channel assignment in wireless sensor networks: a game theoretic approach. In: INFOCOM '10: Proceedings of the 29th IEEE International Conference on Computer Communications, San Diego, CA, 2010
10. Zhou, G., Huang, C.D., Yan, T., He, T., Stankovic, J.A., Adbelzaher, T.F.: "MSN: multi-frequency media access control for wireless sensor networks. In: INFOCOM '06: Proceedings of the 25th IEEE International Conference on Computer Communications, Barcelona, 2006
11. Kim, Y., Shin, H., Cha, H.: Y-MAC: an energy-efficient multi-channel MAC protocol for dense wireless sensor networks. In: IPSN '08: Proceedings of the 7th International Conference on Information Processing in Sensor Networks, St. Louis, MO, 2008
12. ET Docket No. 03-237, Notice of Proposed Rule Making and Order. FCC, 2003

13. Bahl, P., Chandra, R., Moscibroda, T., Murty, R., Welsh, M.: White space networking with Wi-Fi like connectivity. In: SIGCOMM '09: Proceedings of the ACM SIGCOMM Conference on Data Communication, Barcelona, 2009
14. Zhang, L., Zhu, Q., Chen, A.: Fast message dissemination tree and balanced data collection tree for wireless sensor network. J. Softw. **8**(6), 1346–1352 (2013)
15. Ghosh, A., Incel, O.D., Kumar, V.A., Krishnamachari, B.: Multi-channel scheduling for fast aggregated converge cast in wireless sensor networks. In: MASS '09: Proceedings of the 6th IEEE International Conference on Mobile Ad Hoc and Sensor Systems, Macao, 2009
16. Incel, O.D., Krishnamachari, B.: Enhancing the data collection rate of tree-based aggregation in wireless sensor networks. In: SECON '08: Proceedings of the 5th Annual IEEE Communications Society Conference on Sensor, Mesh and Ad Hoc Communications and Networks, San Francisco, CA, 2008
17. Ji, S., Li, Y., Jia, X.: Capacity of dual-radio multi-channel wireless sensor networks for continuous data collection. In: INFOCOM '11: Proceedings of the 30th IEEE International Conference on Computer Communications, Shanghai, 2011
18. Chen, S., Tang, S., Huang, M., Wang, Y.: Capacity of data collection in arbitrary wireless sensor networks. In: INFOCOM '10: Proceedings of the 29th IEEE International Conference on Computer Communications, San Diego, CA, 2010
19. Li, X.-Y., Zhao, J., Wu, Y.W., Tang, S.J., Xu, X.H., Mao, X.F.: Broadcast capacity for wireless ad hoc networks. In: MASS '08: Proceedings of the 5th IEEE International Conference on Mobile Ad Hoc and Sensor Systems, Atlanta, GA, 2008
20. Namboothiri, P.G., Sivalingam, K.M.: Throughput analysis of multiple channel based wireless sensor networks. Wireless Netw. **19**(4), 461–476 (2013)
21. So, H.-S.W., Walrand, J., Mo, J.: McMAC: a parallel rendezvous multi-channel MAC protocol. In: WCNC '07: Proceedings of the IEEE Wireless Communications and Networking Conference, Hong Kong, 2007
22. van Hoesel, L., Havinga, P.: A lightweight medium access protocol (LMAC) for wireless sensor networks. In: INSS '04: Proceedings of the 1st International Conference on Networked Sensing Systems, Tokyo, 2004
23. Halkes, G., Langendoen, K.: Crankshaft: an energy-efficient MAC-protocol for dense wireless sensor networks. In: EWSN '07: Proceedings of the 4th European Conference on Wireless Sensor Networks, Delft, 2007
24. Incel, O.D., Dulman, S., Jansen, P.: Multi-channel support for dense wireless sensor networking. In: EUROSSC '06: Proceedings of the 1st European Conference on Smart Sensing and Context, vol. 4272, pp. 1–14 (2006)
25. Xun, C., Peng, H., Qiu-sheng, H., Shi-liang, T., Zhang-long, C.: A multi-channel MAC protocol for wireless sensor networks. In: CIT '06: Proceedings of the 6th IEEE International Conference on Computer and Information Technology, Seoul, 2006
26. Ansari, J., Zhang, X., Mahonen, P.: Demo abstract: multi-radio medium access control protocol for wireless sensor networks. In: SenSys '07: Proceedings of the 5th International Conference on Embedded Networked Sensor Systems, Sydney, 2007
27. Wang, X.D., Wang, X.R., Fu, X., Xing, G.L., Jha, N.: Flow-based real-time communication in multi-channel wireless sensor networks. In: EWSN '09: Proceedings of the 6th European Conference on Wireless Sensor Networks, Cork, 2009
28. Chipara, O., He, Z., Xing, G., Chen, Q., Wang, X., Lu, C., Stankovic, J., Abdelzaher, T.: Real-time power-aware routing in sensor networks. In: IWQoS '06: Proceedings of the 14th IEEE International Workshop on Quality of Service, New Haven, CT, 2006
29. Moad, S., Hansen, M.T., Jurdak, R., Kusy, B., Bouabdallah, N., Ksentini, A.: On balancing between minimum energy and minimum delay with radio diversity for wireless sensor networks. In: Proceedings Wireless Days, Dublin, 2012
30. Ganesan. D., Govindan, R., Shenker, S., Estrin, D.: Highly-resilient, energy-efficient multipath routing in wireless sensor networks. In: Proceedings of ACM SIGMOBILE Mobile Computing and Communications Review, New York, NY, 2001

31. Pal, A., Nasipuri, A.: DRCS: a distributed routing and channel selection scheme for multi-channel wireless sensor networks. In: PerSeNS '13: Proceedings of the 9th IEEE International Workshop on Sensor Networks and Systems for Pervasive Computing, San Diego, CA, 2013
32. Wu, Y., Keally, M., Zhou, G., Mao, w.: Traffic-aware channel assignment in wireless sensor networks. In: WASA '09: Proceedings of the 4th International Conference on Wireless Algorithms, Systems, and Applications, Boston, MA, 2009
33. Le, H., Henriksson, D., Abdelzaher, T.: A practical multi-channel media access control protocol for wireless sensor networks. In: IPSN '08: Proceedings of the 7th International Conference on Information Processing in Sensor Networks, St. Louis, MO, 2008
34. Yuanyuan, Z., Xiong, N., Park, J.H., Yang, L.T.: An interference-aware multichannel media access control protocol for wireless sensor networks. J. Supercomput. **60**(3), 437–460 (2012)
35. Campbell, C.E.-A., Khan, S., Singh, D., Loo, K.-K.: Multi-channel multi-radio using 802.11 based media access for sink nodes in wireless sensor networks. Sensors **11**(5), 4917–4942 (2011)
36. Canthadai, A.M., Radhakrishnan, S., Sarangan, V.: Multi-radio wireless sensor networks: energy efficient solutions for radio activation. In: GLOBECOM '10: IEEE Global Telecommunications Conference, Miami, FL, 2010
37. Cheng, W., Chen, X., Znati, T., Lu, X., Lu, Z.: The complexity of channel scheduling in multi-radio multi-channel wireless networks. In: INFOCOM '09: Proceedings of the 28th IEEE International Conference on Computer Communications, Rio de Janeiro, 2009
38. Lymberopoulos, D., Priyantha, N.B., Goraczko, M., Zhao, F.: Towards energy efficient design of multi-radio platforms for wireless sensor networks. In: IPSN '08: Proceedings of the 7th International Conference on Information Processing in Sensor Networks, St. Louis, MO, 2008
39. Stathopoulos, T., Lukac, M., McIntire, D., Heidemann, J., Estrin, D., Kaiser, W.J.: End-to-end routing for dual-radio sensor networks. In: INFOCOM '07: Proceedings of the 26th IEEE International Conference on Computer Communications, Anchorage, AK, 2007
40. Gummeson, J., Ganesan, D., Corner, M.D., Shenoy, P.: An adaptive link layer for range diversity in multi-radio mobile sensor networks. In: INFOCOM '09: Proceedings of the 28th IEEE International Conference on Computer Communications, Rio de Janeiro, 2009
41. Irwin, R.E.: Traffic-aware channel assignment for multi-transceiver wireless networks. Ph.D. Dissertation, Virginia Polytechnic Institute and State University, Blacksburg (2012)

Chapter 2
Coverage, Connectivity, and Deployment in Wireless Sensor Networks

Yun Wang, Yanping Zhang, Jiangbo Liu and Rahul Bhandari

Abstract Recent years have witnessed successful real-world deployments of wireless sensor networks (WSNs) in a wide range of civil and military applications. Sensing coverage and network connectivity are two of the most fundamental issues to ensure effective environmental sensing and robust data communication in a WSN application. This chapter presents fundamental studies on the sensing coverage and the network connectivity from mathematical modeling, theoretical analysis, and performance evaluation perspectives. Both lattice WSNs that follow a pattern-based deployment strategy and random WSNs that follow a random deployment strategy are considered. The aim of this chapter is to deliver a systematic study on the fundamental problems in WSNs and provide guidelines in selecting critical network parameters for WSN design and implementation in practice.

Keywords Coverage · Connectivity · Deployment · Wireless sensor networks

2.1 Introduction to Wireless Sensor Network

A wireless sensor network (WSN) is made up of tens to thousands of interconnected sensors that are randomly or deterministically deployed in a field of interest to monitor various environmental changes such as light, temperature, air pressure,

Y. Wang (✉) · J. Liu · R. Bhandari
Department of Computer Science and Information Systems, Bradley University, Peoria, IL, USA
e-mail: ywang2@fsmail.bradley.edu

J. Liu
e-mail: jiangbo@fsmail.bradley.edu

R. Bhandari
e-mail: rbhandari@mail.bradley.edu

Y. Zhang
Department of Computer Science, Gonzaga University, Spokane, WA, USA
e-mail: zhangy@gonzaga.edu

S. Patnaik et al. (eds.), *Recent Development in Wireless Sensor and Ad-hoc Networks*, Signals and Communication Technology, DOI 10.1007/978-81-322-2129-6_2

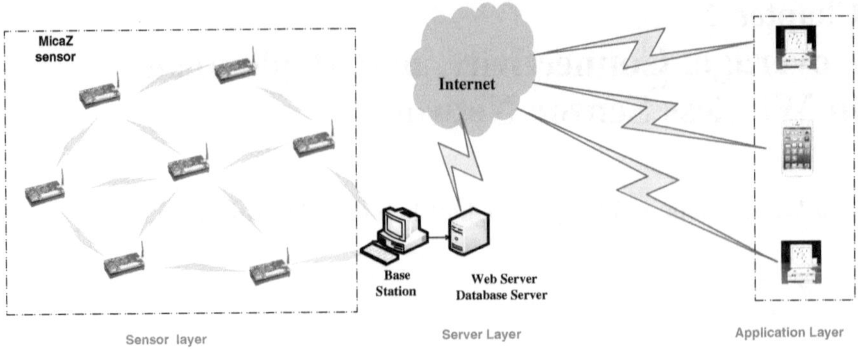

Fig. 2.1 Illustration of a wireless sensor network where sensors periodically sense and detect application-specific events and report the sensing data to a BS for data aggregation, analysis, and visualization

humidity, pollution, etc. Figure 2.1 illustrates a typical WSN where a number of MicaZ[1] sensors are deployed in a square field of interest to monitor several application-specific environmental events. All sensors are connected to a base station (BS) via wireless communication links, and the sensing data are periodically collected and aggregated at the BS. The BS can be connected to a designated Web server that makes it possible for authorized users to remotely access and configure the WSN anytime, anywhere via traditional computers and mobile devices such as iPhone, iPod, iPad, smart phones, etc. MoteLab, for example, is a Web-based sensor network test bed at Harvard University where a set of permanently deployed sensors is connected to a central Web server which provides a Web interface for data accessing and logging, sensor reprogramming, task creating, and scheduling on the test bed [1].

The unit of a WSN is sensor node. Each sensor node has four major components: (1) sensing transducer, (2) data processor, (3) radio transceiver, and (4) embedded batteries [2]. Figure 2.2 illustrates the main components of a MicaZ sensor. The sensing transducer first measures the environmental conditions in its surrounding within certain sensing range and then the data processor transforms the sensed data into an electric signal. After that, the radio transceiver transmits the processed signal to a BS through a direct communication path or a multi-hop communication path for data fusion. All the operations including sensing, computation, and communications are powered by the embedded batteries that are usually non-rechargeable.

[1] Source for MicaZ sensors is available at http://www.openautomation.net/uploadsproductos/micaz_datasheet.pdf.

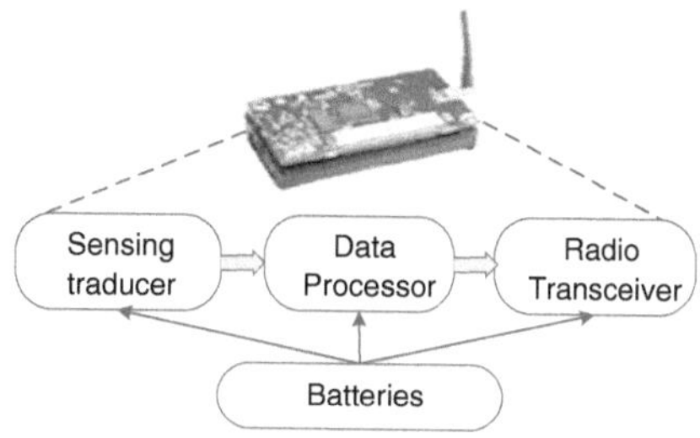

Fig. 2.2 Four major components in a sensor node

2.2 Applications of Wireless Sensor Networks

WSNs are emerging as a new computing platform and networking structure to couple the physical world around us with digital world [3]. It enables novel applications in a wide range of disciplines [4] such as environmental monitoring, habitat monitoring, industrial and manufacturing automation, health care [5], and intrusion detection and tracking. This is due to the fact that consistent sensing data collected by a WSN makes it possible for engineering scientists and engineers to derive quantitative measurements of the dynamics of environmental conditions to either have a better understanding of the monitored field or to capture the occurrence of a set of application-desired events, so as to take appropriate actions whenever needed. Typical applications of WSNs include, but not limited to:

- *Environmental Observation and Forecasting Systems* (EOFS): EOFS is a distributed WSN system designed to monitor, model, and forecast wide-area physical systems such as river systems, transportation, and agriculture for natural resource planning and disaster response [6].
- *Endangered Species Recovery*: To assist the recovery of rare and endangered species of plants, a set of sensors are used to monitor various ecological conditions such as temperature, humidity, rainfall, wind, and solar radiation near endangered plants. Collected data can be used to investigate why a species is rare and to evaluate possible remedial actions [7].
- *Habitat Monitoring* [8, 9]: An application of WSN to monitor the habitat of sensible wildlife through sampling the environmental changes in terms of temperature, humidity, barometric pressures, and midrange. As an example, in 2002, about three dozen UC Berkeley Motes were deployed on Great Duck Island, Maine, to monitor the microclimates in nesting burrows used by the Leach's storm petrel and study the habitat of storm petrel [10].
- *Intrusion Detection and Tracking*: Detection, classification, and tracking of intruders/targets/objects are a basic surveillance or military application of WSN [11] and have been studied in the literature from many aspects [11–18]. Such applications concern how fast the WSN can detect certain intruders/targets/objects and how reliably the sensing and detection data can be reported to the BS.
- *Structural and Seismic Monitoring* [19–21]: An application of WSN in the field of civil engineering to monitor the condition of civil structures such as

buildings, bridges, roads, and aircrafts for instrumentation. It is considered as substitutes for traditional tethered monitoring systems due to its low cost, ease of deployment, and lack of wiring.

2.3 Preliminaries

In a WSN, every sensor has a limited sensing range, denoted as r_s, and a limited communication range, denoted as r_c. The union of the sensing ranges of all sensors is defined as the network *sensing coverage* [22], which reflects how well the area of sensor field is monitored. In addition, to communicate successfully, a WSN must provide satisfactory *network connectivity*, so as to eliminate the isolation of sensors and enable each sensor to report its sensing data to its fusion center [23]. In order to understand the sensing coverage and network connectivity in a WSN, several fundamental models including network deployment model, sensing model, and communication model must be introduced.

2.3.1 Network Deployment Model

According to the accessibility of the monitored area, there are two primary sensor deployment strategies: (1) *deterministic sensor deployment* in a controlled and human-friendly environment [24] and (2) *random sensor deployment* in a dangerous and inaccessible region [25]. Generally speaking, the pattern-based lattice WSNs resulting from deterministic sensor deployment provide better sensing coverage and higher degree of connectivity [26], comparing to the random counterparts. On the other hand, it was found in [27, 28] that deterministic sensor deployment does not always outperform the random sensor deployment for applications that do not require full coverage such as intrusion detection.

Assume that the field of interest is a two-dimensional square region with area $A = L \times L$. A number of N sensors are deployed in the target region A, and the location of sensor i is represented by (x_i, y_i), for $i = 1,\ldots, N$, over the two-dimensional region. In a lattice WSN, the sensors' locations conform to the geographical pattern shapes. On the other hand, in a WSN following a random and uniform sensor deployment, the probability density function (PDF) for a sensor at location (x_i, y_i) is given by $f(x_i, y_i) = \frac{1}{A}$.

Figures 2.3 and 2.4 illustrate a square pattern-based lattice WSN and a randomly and uniformly distributed WSN, respectively, where 100 sensors are deployed in a square region of interest with area $A = 100 \times 100$. In addition, the Voronoi diagrams of both WSNs are depicted [29].

Voronoi diagram of a WSN presents the proximity information about the deployment sensors in the field of interest [30]. To be specific, the Voronoi diagram

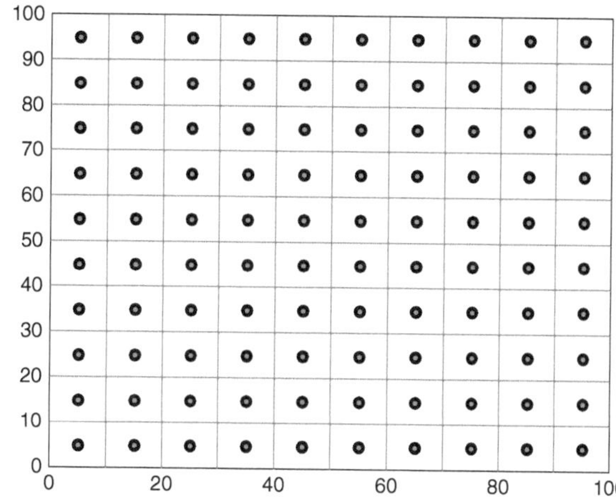

Fig. 2.3 Voronoi diagram of a square pattern-based lattice WSN

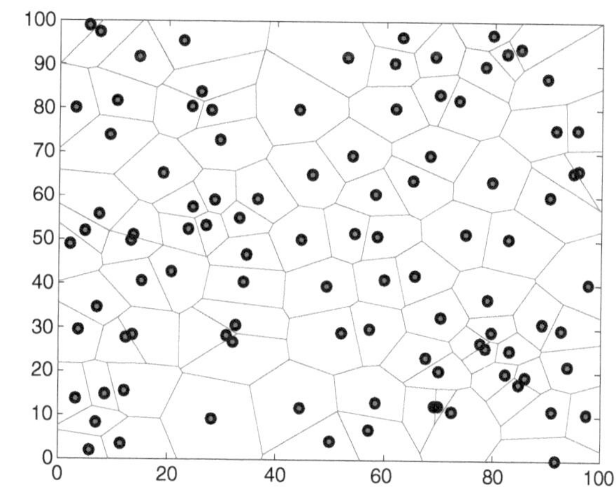

Fig. 2.4 Voronoi diagram of a random WSN

of the N sensors partitions the field of interest into N Voronoi polygons. Each sensors has a Voronoi polygon, where all the inside points are closer to it than to any other sensors [31, 32]. It is clear that the Voronoi polygon for each sensor is the same for all sensors in a lattice WSN. This property can be exploited to analyze the deployment efficiency of lattice WSNs [29, 30, 33], to be discussed in Sect. 2.4.

2.3.2 Network Connectivity

The communication typology of a WSN can be modeled as a graph, denoted by $G = (V, E)$, where V is the set of sensor vertices and E is the set of wireless communication links, i.e., the line segments connecting neighboring sensors. A pair of sensors is said to be *neighbors* of each other if their Euclidean distance is at most the communication range r_c, according to the *disk communication model*. As illustrated

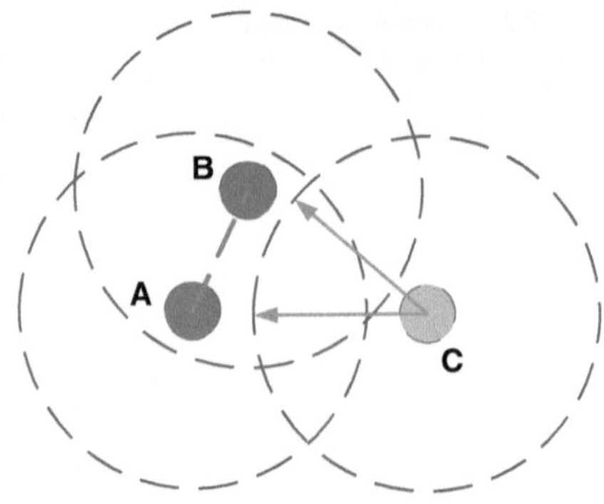

Fig. 2.5 Communication links in a WSN graph

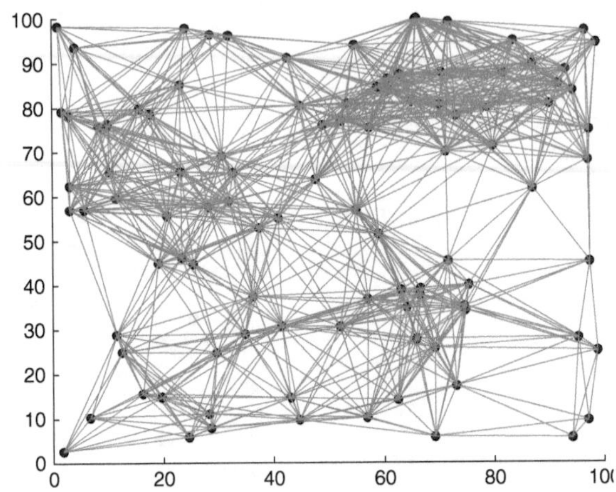

Fig. 2.6 Illustration of a connected WSN graph where 100 sensors are randomly and uniformly deployed in a square region with area $100 \times 100\ \mathrm{m}^2$, the communication range is set as 30 m

in Fig. 2.5, sensors A and B are neighbors and there is a bi-directional link between them as they are within each other's communication range while sensor C is isolated.

A WSN is defined as *connected* if, for any two sensors, there is a single-hop or multi-hop communication path between them consisting of consecutive wireless communication links. The WSN connectivity is primarily determined by the sensors' deployment locations and the communication ranges. Figure 2.6, for example, illustrates a *connected* graph of a WSN, where 100 sensors are randomly deployed at a square region with area $100 \times 100\ \mathrm{m}^2$ and the communication range of each sensor is set at 30 m. As a contrast, Fig. 2.7 depicts an *un-connected* graph for the WSN when the sensor's communication is reduced to 15 m.

Note that the disk communication model is widely adopted in the state-of-the-art study on the network connectivity of a WSN; however, it is challenged by empirical measurements where the wireless links are found to be highly irregular and far from being isotropic due to multi-path and shadowing effect as well as the environmental noises and interferences [34]. The "signal-to-interference-plus-noise' ratio (SINR) model and the log-normal shadowing model can be adopted to capture these effects as described in [35].

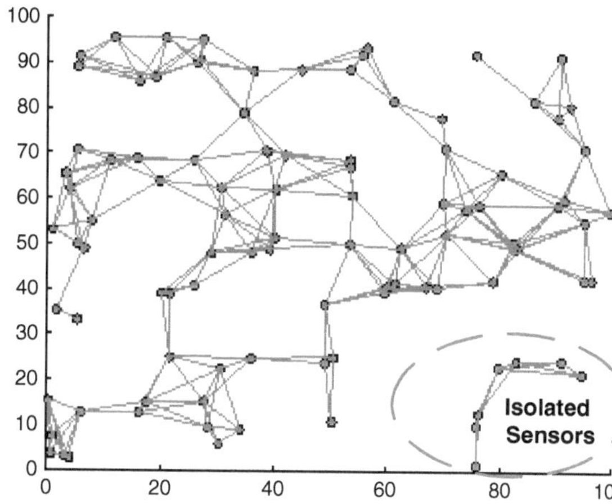

Fig. 2.7 Illustration of an *unconnected* WSN graph where 100 sensors are randomly and uniformly deployed in a square field of 100 × 100 m^2, the communication range is set as 15 meters

2.3.3 Network Coverage

A WSN is deployed in a region of interest to monitor certain environmental conditions or changes in an application. A point is covered by a sensor if the Euclidean distance between the point and the sensor is no more than the sensing range r_s according to the *disk/Boolean sensing model* [36, 37]. Disk sensing model is widely used due to its simplicity and the fact that it enables theoretical abstraction and analysis. Mathematically speaking, a point p is covered by a sensor s_i, if and only if,

$$d(s_i, p) \le r_c, \quad (2.1)$$

where $d(s_i, p)$ is the Euclidean distance between point p and the sensor s_i.

For *full sensing coverage*, the entire region of application field is covered by at least one sensor and there is no sensing coverage hole in the network. Figures 2.8

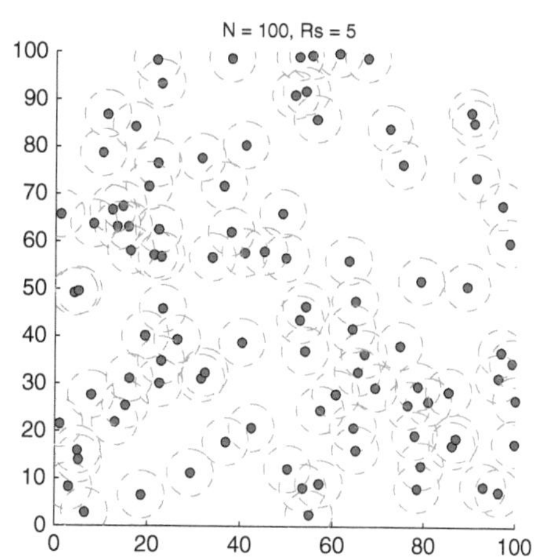

Fig. 2.8 Sensing coverage of a random WSN with 100 sensors and sensing range *R5* = 5 m. The field of interest is *not fully* covered

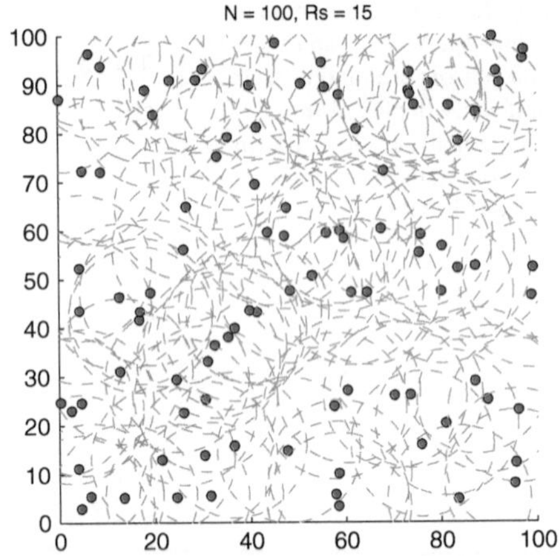

Fig. 2.9 Sensing coverage of a random WSN with 100 sensors and sensing range *R*5 = 15 m. The field of interest is nearly fully covered

and 2.9 represent the sensing coverage of a random WSN with 100 sensors for sensing range $r_s = 5$ and $r_s = 15\,\text{m}$, respectively.

The disk sensing model may not be able to simulate the sensing capability of a real-life sensor accurately, because realistic sensors are designed to be small and cheap that they are unlikely to be sophisticated enough to provide exactly the same detection capability in every direction. A probabilistic model, *Elfes sensing model* [38, 39], is therefore used in the literature to better model the realistic sensing capability of a real-life sensor.

Following the Elfes sensing model, the probability that a sensor detects an event to a distance d is [38, 39]:

$$P(d) = \begin{cases} 1, & d < (r_s - e); \\ e^{-a(r-r_s)^{\gamma}}, & (r_s - e) \le d \le (r_s + e); \\ 0 & d > (r_s + e); \end{cases} \tag{2.2}$$

where r_s is the average sensing range, e is a measure of the sensing uncertainty in sensor's detection capability, and α and γ are device-oriented parameters. To be specific, a sensor can detect an event at a distance less than r_s - e with probability 1, at a distance greater than $r_s + e$ with probability 0, and in the distance interval $(r_s - e, r_s + e)$ with probability $e^{-a(r-r_s)^{\gamma}}$ [35, 40]. Figure 2.10 depicts the sensing and detection probability of a sensor under the Elfes sensing model and the Boolean sensing model when $r_s = 5$ and $e = 3$.

Note that Boolean sensing model can be regarded as a special case of Elfes sensing model by setting $e = 0$. A study on the impacts of sensing models on the sensing coverage of a WSN is presented in [39].

Fig. 2.10 Elfes sensing model

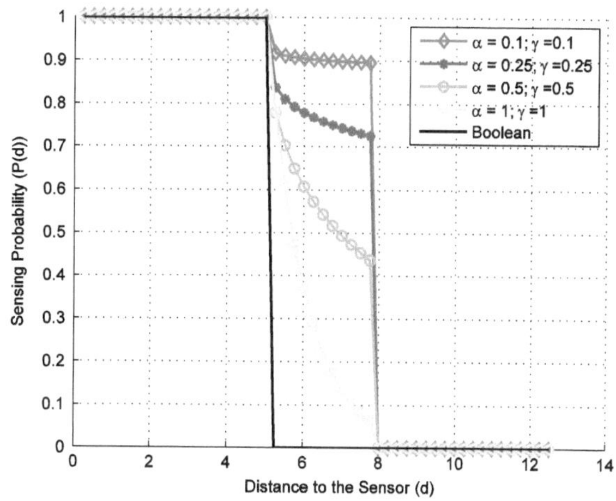

2.4 Coverage and Connectivity in Lattice WSNs

In lattice WSNs, sensors are deliberately and precisely placed at desirable locations in a controlled fashion following regular patterns. Popular patterns including *square, triangle, and hexagon* that can be repeated to cover a continuous region without having any overlapping areas are widely adopted in practice due to the simplicity and the convenience of deployment. Other patterns such as rhombus and mutated patterns [30, 41, 42, 43, 44] are also investigated in the literature in terms of minimizing the required number of sensors while providing desirable quality of service (QoS) for various application contexts. Here, we focus on the three popular patterns and compare their efficiency in providing *full sensing coverage*, *full network connectivity*, and *full sensing coverage with connectivity* under the same application context.

Figure 2.11 illustrates the Voronoi polygon area for a sensor in a lattice WSN following a square, a triangular, and a hexagonal pattern, respectively. It was pointed out in [32] that the area of the Voronoi polygon for a sensor represents the average contribution to the network QoS such as sensing coverage and communication

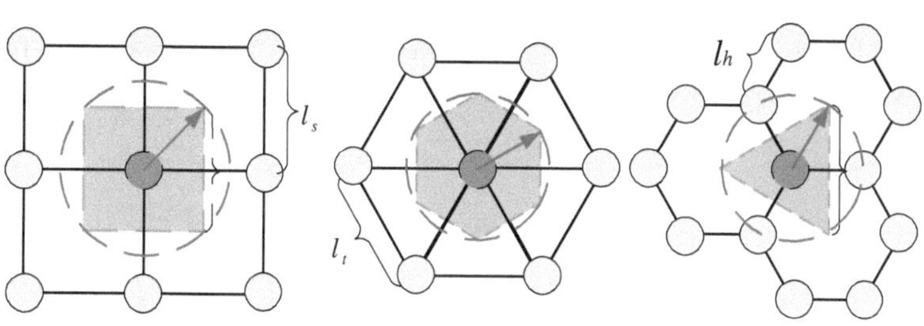

Fig. 2.11 Voronoi polygon for a sensor in a lattice WSN following a *square*, a *triangular*, and a *hexagonal* pattern, respectively [29]

connectivity. It also determines the required number of sensors. To be specific, given an application field with area A and assuming A_s is the area of the Voronoi polygon for a sensor in a lattice WSN, the number of deployed sensors can be estimated as:

$$N = \frac{A}{A_s}. \tag{2.3}$$

The area of the Voronoi polygon for a sensor is determined by the side length of the deployment pattern, i.e., the deployment distance l_x. The deployment distance is determined by the sensor's capability of sensing range r_s, communication range r_c, and the application requirements on sensing coverage and connectivity.

2.4.1 Full Connectivity in Lattice WSNs

For *full connectivity*, the side length of the deployment pattern should be no more than the communication range, i.e., $l_x \leq r_c$. Optimally, the side length of the deployment pattern is set as the communication range r_c for maximum deployment efficiency in that minimum sensors are needed. Specifically, to provide full connectivity, the maximum area of the Voronoi polygon for each sensor is given as [29]:

$$A_s(\text{Con}) = \begin{cases} l_s^2 = r_c^2, & \text{square} \\ \frac{\sqrt{3}}{2} l_t^2 = \frac{\sqrt{3}}{2} r_c^2, & \text{triangular} \\ \frac{3\sqrt{3}}{4} l_h^2 = \frac{3\sqrt{3}}{4} r_c^2, & \text{hexagonal} \end{cases} \tag{2.4}$$

for a square, a triangular, and a hexagonal pattern-based WSN, respectively.

Combining Eqs. (2.3) and (2.4), the required number of sensors to achieve full connectivity in a square field of interest with area $A = 100 \times 100\ \text{m}^2$ is plotted in Fig. 2.12 for the three considered patterns. The communication range is varied from 3.25 to 14 m, and the required number of sensors changes from about 900–100 in the studied cases. It can be observed that *hexagonal* pattern is the optimal deployment pattern in terms of needing the minimum number of sensors as compared to the square and triangular patterns for full connectivity. Note that 4-connectivity, 6-connectivity, and 3-connectivity are automatically provided in fully connected lattice WSN following a square, a triangular, and a hexagonal pattern, respectively.

2.4.2 Full Coverage in Lattice WSNs

On the other hand, for full sensing coverage, the maximum side length of the Voronoi polygon for a sensor should be $\sqrt{2}r_s$, r_s, and $\sqrt{3}r_s$ for a square, a triangular,

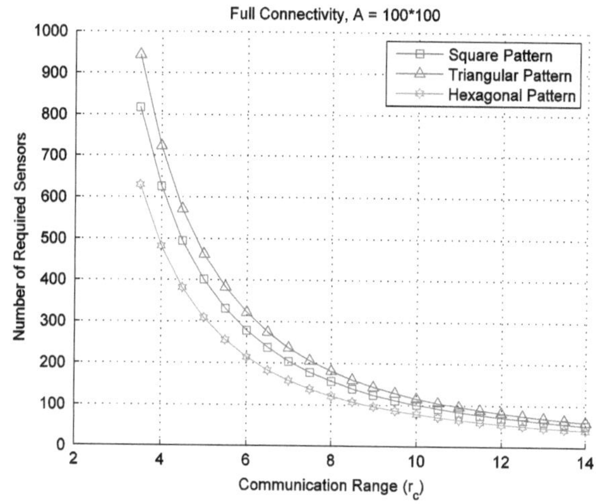

Fig. 2.12 Number of required sensors for full connectivity in a lattice WSN following a *square*, a *triangular*, and a *hexagonal* pattern, respectively

and a hexagonal pattern-based lattice WSN, respectively, based on the geometry illustrated in Fig. 2.11. The maximum area of the Voronoi polygon is thus computed as [29]:

$$A_5(\text{Cov}) = \begin{cases} 2r_s^2, & \text{square} \\ \frac{3\sqrt{3}}{2}r_s^2, & \text{triangular} \\ \frac{3\sqrt{3}}{4}r_s^2, & \text{hexagonal} \end{cases} \tag{2.5}$$

for a square, a triangular, and a hexagonal pattern-based WSN, respectively.

Similarily, combining Eqs. (2.3) and (2.5), we have the required number of sensors to achieve full sensing coverage in a square field of interest with area $A = 100 \times 100\ \text{m}^2$ for the three considered patterns as illustrated in Fig. 2.13. In this figure, the sensing range is varied from 2 to 10 m. It can be observed that *triangular* pattern is the optimal deployment pattern in terms of needing the minimum number of sensors as compared to square and hexagonal patterns for full sensing coverage.

2.4.3 Full Coverage and Connectivity in Lattice WSNs

To achieve both full sensing coverage and full connectivity in a WSN where each point in the field of interest should be covered and all sensors in the network is connected; the above analysis on the sensing coverage and the network connectivity should be combined. To be specific, Eqs. (2.4) and (2.5) give the maximal area of the Voronoi polygon $A_5(\text{Con})$ and $A_5(\text{Cov})$ for a given sensor in the three considered lattice WSNs for full connectivity and full sensing coverage, respectively. The smaller one, i.e., $\min(A_5(\text{Con}), A_5(\text{Cov}))$, is the design bottleneck and

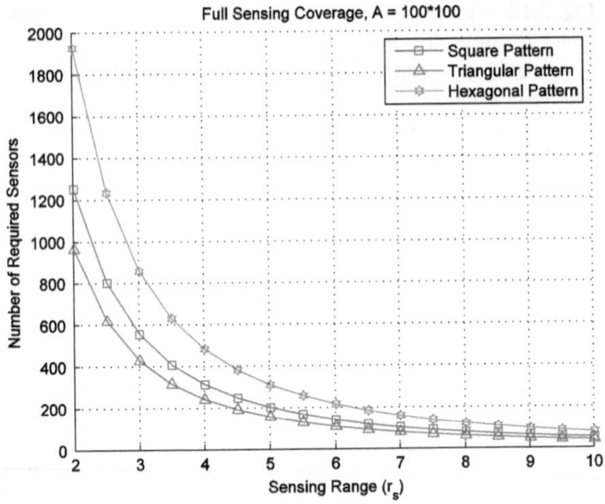

Fig. 2.13 Required number of sensors for full sensing coverage in a lattice WSN following *square*, *triangular*, and *hexagonal* patterns

determines the maximal area of the Voronoi polygon of a sensor for both full sensing coverage and full connectivity and is given by [30]:

$$A_s(\mathrm{Cov}, \mathrm{Con}) = \begin{cases} \min(2r_s^2, r_c^2), & \text{square} \\ \min\left(\frac{3\sqrt{3}}{2}r_s^2, \frac{\sqrt{3}}{2}r_c^2\right), & \text{triangular} \\ \min\left(\frac{3\sqrt{3}}{4}r_s^2, \frac{3\sqrt{3}}{4}r_c^2\right), & \text{hexagonal} \end{cases} \quad (2.6)$$

for a square, a triangular, and a hexagonal pattern-based WSN, respectively.

Clearly, the ratio of the sensing range to the communication range plays an important impact on the area of the unit Voronoi polygon for each sensor and the deployment efficiency. Combining Eqs. (2.3) and (2.6), we have the required number of sensors to achieve both full sensing coverage and full connectivity in a region of interest with area $A = 100 \times 100\ \mathrm{m}^2$ for the three considered patterns as illustrated in Fig. 2.14. In this analysis, the communication range is fixed at 5 and the sensing range is varied from 2.25 to 7.25 to generate different ratios of sensing range to communication ranges (varies from 0.45 to 1.45). It can be observed in the figure that the ratio of sensing range to communication range determines the optimal deployment pattern for the minimum required number of sensors and no pattern is always the best for all the cases.

Specifically, triangular pattern is the optimal one when $r = \frac{r_s}{r_c} \leq 0.6580$, square pattern outperforms the other two patterns when $0.6580 < r = \frac{r_s}{r_c} < 0.8772$, and hexagonal pattern is the best pattern when $r = \frac{r_s}{r_c} \geq 0.87725$. Bai et al. discussed more patterns and their efficiencies in [30, 41, 42, 43, 44].

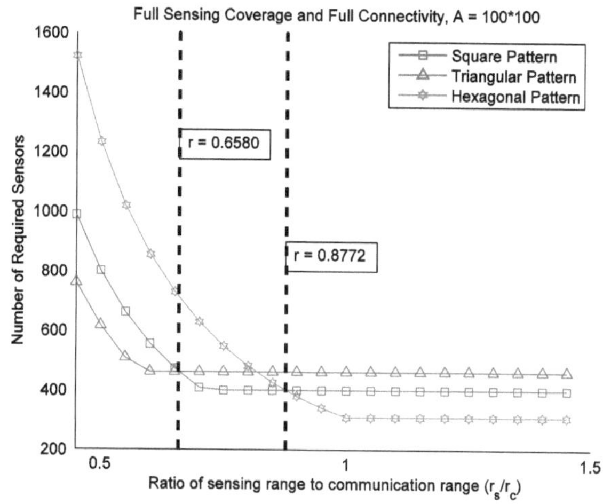

Fig. 2.14 Required number of sensors for both full sensing coverage and full connectivity in a lattice WSN following a *square*, a *triangular*, and a *hexagonal* pattern

2.5 Coverage and Connectivity in Random WSNs

Random deployment is appealing for large-scale WSN applications, where sensors can be dropped from a low-flying airplane or an unmanned aerial vehicle (UAV). This deployment strategy is easily scalable and is appropriate for a hostile environment in many civil and military applications such as contaminant transport monitoring and intrusion detection [45]. The sensing coverage and network connectivity based on a random WSN model have been extensively investigated in the literature [46–57].

2.5.1 Coverage of Random WSNs

In a random WSN where a number of N sensors are deployed independently and uniformly in a bounded field of interest A as illustrated in Fig. 2.15, the node density is $p = N/A$. Assuming the disk sensing model where each sensor can monitor any point whose distance to the sensor is no more than the sensing range r_s, the network sensing coverage is derived as [37, 53, 58]:

$$P\text{cov} = 1 - e^{-p\pi r_s^2} \tag{2.7}$$

Consider an arbitrary point T, the probability that there exist n sensors in the circular area S_T centered at T with radius r_s is Poisson distributed and can be derived as:

$$P(n, T) = \frac{(pS_T)^n}{n!} e^{-pS_T}, \tag{2.8}$$

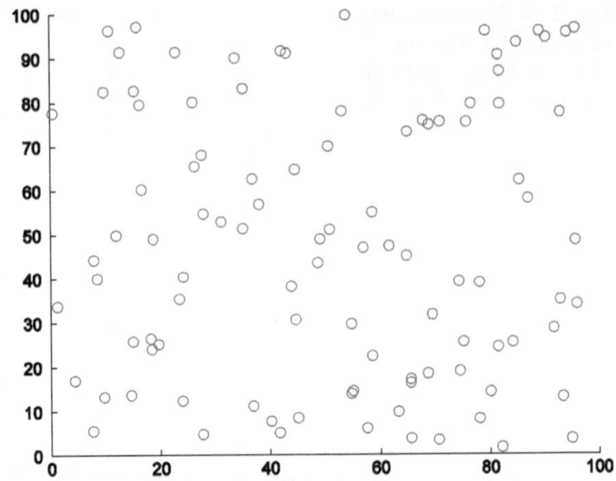

Fig. 2.15 An example random WSN with 100 independently and uniformly deployed sensors

where $S_T = \pi r_s^2$. The probability that there is no sensor (i.e., $n = 0$) deployed in S_T is derived as:

$$P(0,T) = e^{-p\pi r_s^2}. \tag{2.9}$$

Hence, the probability that there is at least one sensor deployed in the S_T (the point T is covered by at least one nearby sensors) can be computed as:

$$P\text{cov}(T) = 1 - e^{-p\pi r_s^2}. \tag{2.10}$$

As the N sensors are independently and uniformly distributed over the entire field of interest, all points in the field of interest have identical probability of being covered. Thus, the network sensing coverage is:

$$P\text{cov} = Pcv(T) = 1 - e^{-p\pi r_s^2}. \tag{2.11}$$

Figure 2.16 illustrates the numerical results of the network sensing coverage versus the sensing range for different numbers of deployed sensors, i.e., $N = 100$, 200, 300, 400, and 500, respectively. It is clear that the network sensing coverage is jointly determined by the network node density ($p = N/A$) and the sensors' sensing range r_s. Increasing sensors' sensing range and/or increasing the node density can be used to increase the network sensing coverage. On the other hand, given an application-specific sensing coverage requirement *P*cov and sensors' sensing range r_s, the required number of sensors can be determined according to Eq. (2.11). A recent survey on the Coverage Problems in Sensor Networks is provided in [59].

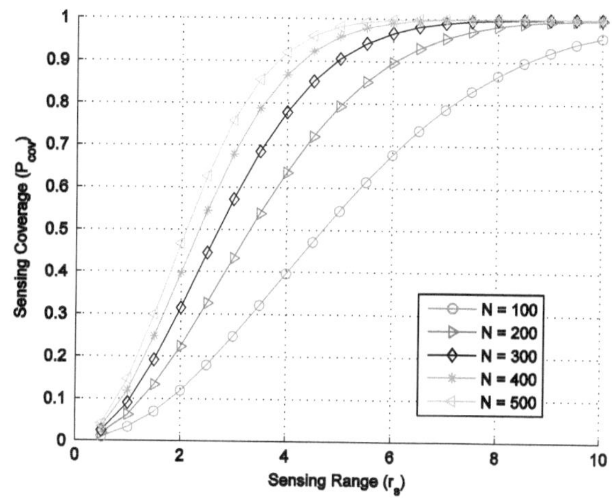

Fig. 2.16 Probability that any point in the WSN field of interest is covered by at least one nearby sensor(s), i.e., the sensing 1-coverage of a random WSN with respect to the sensing range

2.5.2 Connectivity of Random WSNs

Network connectivity is another fundamental problem in WSNs to ensure the data communication among sensors and the BS. A sensor is said to be connected if and only if it has a direct or multi-hop communication path to the BS. A WSN is said to be connected if all sensors are connected. Figures 2.6 and 2.7 illustrate a connected WSN and un-connected WSN, respectively. It is clear that in a connected WSN, there is no isolated sensor(s) in the network. A sensor is said to be isolated if its node degree is zero. The node degree of a sensor is the number of neighbors within its communication range r_c. As illustrated in Fig. 2.5, node *C* is an isolated sensor and has no neighbor within its communication range.

Consider an arbitrary sensor s_i in the network, the probability that it is isolated is equivalent to the probability that there exists no neighboring sensor within its communication range and follows:

$$P_{iso}(s_i) = e^{-p\pi r_c^2}. \tag{2.12}$$

The probability that sensor s_i is *not* isolated is:

$$P_{\text{non-iso}}(s_i) = 1 - e^{-p\pi r_c^2}. \tag{2.13}$$

As all *N* sensors are independently and uniformly deployed in the field of interest, they have the same probability of being non-isolated. Thus, the probability that there is no isolated sensor in the WSN is given as:

$$P_{\text{non-iso}} = \prod_{i=1}^{N} P_{\text{non-iso}}(s_i) = \left(1 - e^{-p\pi r_c^2}\right)^N. \tag{2.14}$$

The event that there is no isolated sensor in the WSN is a necessary (but not sufficient) condition for a WSN to be connected. It gives the *upper bound* of the network connectivity. Further, it is found in [50–52] that a geometric random graph resulting from a random WSN that attains the property that all sensors have at least K neighbors is asymptotically equal to the property that the WSN has K-connectivity [19]. In other words, a WSN in which all sensors have at least one neighbor (i.e., there is no isolated sensor) implies that the WSN is connected with a high probability, i.e.,

$$P_{\text{con}} = P_{\text{non-iso}} = \left(1 - e^{-p\pi r_c^2}\right)^N, \tag{2.15}$$

for $P_{\text{non-iso}}$ is almost one [51].

Figure 2.17 shows the probability that the WSN is connected with respect to the communication range (r_c) for different numbers of deployed sensors in the square field of interest with area A = 100 × 100 m^2. It can be observed that there exists a critical communication range above which there is a high probability that the WSN is connected. The critical communication range is jointly determined by the sensor distribution and the communication range.

Note that for a WSN to function well, both sensing coverage and connectivity should be ensured. One of the fundamental results on integrated sensing coverage and connectivity is stated as: *If the communication radius of sensors is at least twice the sensing range, i.e.,* $r_c = 2\ r_s$, *then the sensing 1-coverage of a bounded field of interest is sufficiently to guarantee 1-connectivity of the WSN* [35, 48, 49].

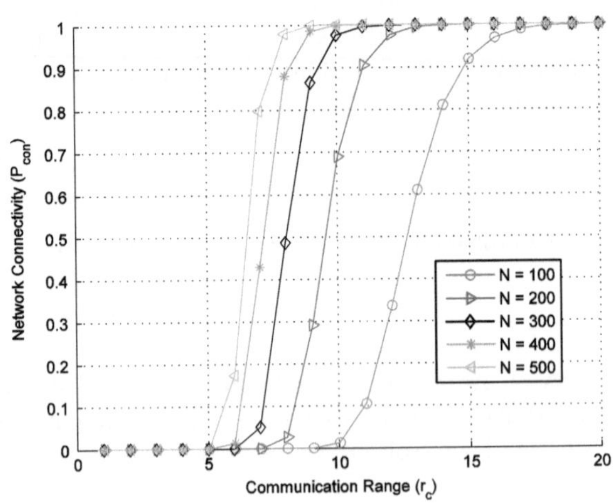

Fig. 2.17 Probability that a random WSN is connected with respect to the communication range for different numbers of deployed sensors in square field of interest with area A = 100 × 100 m^2

2.6 Summary

In this chapter, we have introduced three fundamental problems: sensing coverage, network connectivity, and sensor placement/deployment in a WSN. Sensing coverage reflecting how well the WSN field of interest is covered and communication connectivity describing how reliable the sensing information can be gathered at the BS are the two of the most basic QoS requirements in a WSN. We have presented the analysis on sensing coverage, connectivity, and connected coverage for lattice WSNs following pattern-based deployment strategy and random WSNs following a random deployment strategy, respectively. In literature, the open problems in this direction include sensing coverage and connectivity analysis in three-dimensional WSNs, non-uniformly distributed WSNs, and mobile WSNs.

References

1. Werner-Allen, G., Swieskowski, P., Welsh, M.: Motelab: A wireless sensor network testbed. In: Proceedings of the 4th International Symposium on Information Processing in Sensor Networks. IEEE Press (2005)
2. Al-Karaki, Jamal N., Kamal, Ahmed E.: Routing techniques in wireless sensor networks: a survey. IEEE Wireless Commun. **11**(6), 6–28 (2004)
3. Zhao, F.: Wireless sensor networks: A new computing platform for tomorrow's internet. In: Proceedings of the IEEE 6th Circuits and Systems Symposium on Emerging Technologies: Frontiers of Mobile and Wireless Communication, vol. 1. pp. I–27. IEEE (2004)
4. Sohraby, K., Minoli, D., Znati, T.: Wireless Sensor Networks: Technology, Protocols, and Applications. John Wiley and Sons Inc., New Jersey (2007)
5. Nakamura, E.F., Loureiro, A.A.F., Frery, A.C.: Information fusion for wireless sensor networks: Methods, models, and classifications. ACM Comput. Surv. (CSUR) **39**(3), 9 (2007)
6. Xu, Ning: A survey of sensor network applications. IEEE Commun. Mag. **40**(8), 102–114 (2002)
7. Biagioni, E., Bridges, K.: The application of remote sensor technology to assist the recovery of rare and endangered species. Special Issue on Distributed Sensor Networks for the International Journal of High Performance Computing Applications, vol. 16(3) (2002)
8. Mainwaring, A., Culler, D., Polastre, J., Szewczyk, R., Anderson, J.: Wireless sensor networks for habitat monitoring. In: Proceedings of the 1st ACM International Workshop on Wireless Sensor Networks and Applications, pp. 88–97. ACM (2002)
9. Szewczyk, R., Osterweil, E., Polastre, J., Hamilton, M., Mainwaring, A., Estrin, D.: Habitat monitoring with sensor networks. Commun. ACM **47**(6), 34–40 (2004)
10. Mainwaring, A., Culler, D., Polastre, J., Szewczyk, R.M., Anderson, J.: Wireless sensor networks for habitat monitoring. In: Proceedings of the 1st ACM International Workshop on Wireless Sensor Networks and Applications, WSNA '02, pp. 88–97 (2002)
11. Arora, A., Dutta, P., Bapat, S., Kulathumani, V., Zhang, H., Naik, V., Mittal, V., Cao, H., Demirbas, M., Gouda, M., et al.: A line in the sand: a wireless sensor network for target detection, classification, and tracking. Comput. Netw. **46**(5), 605–634 (2004)
12. Clouqueur, T., Phipatanasuphorn, V., Ramanathan, P., Saluja, K.K.: Sensor deployment strategy for target detection. In: Proceedings of the 1st ACM International Workshop on Wireless Sensor Networks and Applications, pp. 42–48. ACM (2002)

13. Cao, Q., Yan, T., Stankovic, J., Abdelzaher, T.: Analysis of target detection performance for wireless sensor networks. In: Distributed Computing in Sensor Systems, pp. 276–292. Springer, New York (2005)
14. Cordeiro, C., Agrawal, D.: Ad Hoc and Sensor Networks: Theory and Applications. World Scientific Pub Co Inc., Singapore (2006)
15. Wang, Y., Wang, X., Xie, B., Wang, D., Agrawal, D.P.: Intrusion detection in homogeneous and heterogeneous wireless sensor networks. IEEE Trans. Mob. Comput. **7**(6), 698–711 (2008)
16. Zhu, M., Ding, S., Wu, Q., Brooks, R.R., Rao, N.S., Iyengar, S.S.: Fusion of threshold rules for target detection in wireless sensor networks. ACM Trans. Sensor Netw. (TOSN) **6**(2), 18 (2010)
17. Yick, J., Mukherjee, B., Ghosal, D.: Wireless sensor network survey. Comput. Netw. **52**(12), 2292–2330 (2008)
18. Lin, K.W., Hsieh, M.-H., Tseng, V.S.: A novel prediction-based strategy for object tracking in sensor networks by mining seamless temporal movement patterns. Expert Syst. Appl. **37**(4), 2799–2807 (2010)
19. Krishnamachari, B.: Networking Wireless Sensors. Cambridge University Press, Cambridge (2005)
20. Kim, S., Pakzad, S., Culler, D., Demmel, J., Fenves, G., Glaser, S., Turon, M.: Health monitoring of civil infrastructures using wireless sensor networks. In: 6th International Symposium on Information Processing in Sensor Networks (IPSN 2007), pp. 254–263. IEEE (2007)
21. Lynch, J.P., Loh, K.J.: A summary review of wireless sensors and sensor networks for structural health monitoring. Shock Vib. Dig. **38**(2), 91–130 (2006)
22. Wang, Y., Wang, X., Agrawal, D.P., Minai, A.A.: Impact of heterogeneity on coverage and broadcast reachability in wireless sensor networks. In: The Fifteenth International Conference on Computer Communications and Networks (ICCCN) (2006)
23. Wang, X., Xing, G., Zhang, Y., Lu, C., Pless, R., Gill, C.: Integrated coverage and connectivity configuration in wireless sensor networks. In: ACM SenSys'03 (2003)
24. de Morais Cordeiro, C., Agrawal, D.P.: Ad Hoc and Sensor Networks: Theory and Applications. World Scientific, Singapore (2011)
25. Chakrabarty, K., Iyengar, S.S., Qi, H., Cho, E.: Grid coverage for surveillance and target location in distributed sensor networks. IEEE Trans. Comput. **51**(12), 1448–1453 (2002)
26. Zhang, H., Hou, J.C.: Is deterministic deployment worse than random deployment for wireless sensor networks? In: INFOCOM (2006)
27. Wang, Y., Kutta, A.: Joint and simultaneous K-sensing detection in deterministic and random sensor networks. In: Proceedings of the 26th IEEE International Parallel & Distributed Processing Symposium Workshop on Dependable Parallel, Distributed and Network-Centric Systems. May 21–25, Shanghai, China (2012)
28. Wang, Y., Chu, W., Zhang, Y., Li, X.: Partial sensing coverage with connectivity in lattice wireless sensor networks. Int. J. Sens. Net. **14**(4), 226–240 (2013)
29. Wang, Y., Agrawal, D.P.: Optimizing sensor networks for autonomous unmanned ground vehicles. In: SPIE Symposium on Optics/Photonics in Security & Defence, Cardiff, Wales, United Kingdom, pp. 1–11 (2008)
30. Bai, X., Kumar, S., Xuan, D., Yun, Z., Lai, T.H.: Deploying wireless sensors to achieve both coverage and connectivity. In: Proceedings of the 7th ACM International Symposium on Mobile Ad Hoc Networking and Computing, pp. 131–142. ACM (2006)
31. Rourke, J.: Computational Geometry in C. Cambridge University Press, Cambridge (1998)
32. Fortune, Steven: Voronoi diagrams and Delaunay triangulations. Comput. Euclidean Geom. **1**, 193–233 (1992)
33. Alam, S. M., Haas, Z.J.: Coverage and connectivity in three-dimensional networks. In: Proceedings of the 12th annual international conference on Mobile computing and networking. ACM (2006)

34. Zuniga, M., Krishnamachari, B.: Analyzing the transitional region in low power wireless links. In: First Annual IEEE Communications Society Conference on Sensor and Ad Hoc Communications and Networks (IEEE SECON 2004), pp. 517–526. IEEE (2004)
35. Ghosh, A., Das, S.K.: Coverage and connectivity issues in wireless sensor networks: A survey. Pervasive Mob. Comput. **4**(3), 303–334 (2008)
36. Liu, B., Towsley, D.: A study of the coverage of large-scale sensor networks. In: IEEE International Conference on Mobile Ad-Hoc and Sensor Systems, pp. 475–483. IEEE (2004)
37. Liu, B., Brass, P., Dousse, O., Nain, P., Towsley, D.: Mobility improves coverage of sensor networks. In: Proceedings of the 6th ACM International Symposium on Mobile Ad Hoc Networking and Computing, pp. 300–308. ACM (2005)
38. Elfes, A.: Occupancy grids: a stochastic spatial representation for active robot perception. In: Iyenger, S.S., Elfes, A. (eds.) Autonomous Mobile Robots: Perception, Mapping and Navigation, vol. 1, pp. 60–70. IEEE Computer Society Press (1991)
39. Hossain, A., Biswas, P.K., Chakrabarti, S.: Sensing models and its impact on network coverage in wireless sensor network. In: IEEE Region 10 and the Third International Conference on Industrial and Information Systems (ICIIS 2008), pp. 1–5. IEEE (2008)
40. Wang, Y., Leow, Y.K., Yin, J.: A novel sine-curve mobility model for intrusion detection in wireless sensor networks. Wireless Commun. Mob. Comput. **13**(17), 1555–1570 (2011)
41. Bai, X., Yun, Z., Xuan, D., Jia, W., Zhao, W.: Pattern mutation in wireless sensor deployment. In: INFOCOM, 2010 Proceedings IEEE, pp. 1–9. IEEE (2010)
42. Bai, X., Xuan, D., Yun, Z., Lai, T.H., Jia, W.: Complete optimal deployment patterns for full-coverage and k-connectivity ($k \leq 6$) wireless sensor networks. In: Proceedings of the 9th ACM International Symposium on Mobile Ad Hoc Networking and Computing, pp. 401–410. ACM (2008)
43. Yun, Z., Bai, X., Xuan, D., Lai, T.H., Jia, W.: Optimal deployment patterns for full coverage and k-connectivity ($k \leq 6$) wireless sensor networks. IEEE/ACM Trans. Netw. (TON) **18**(3), 934–947 (2010)
44. Bai, X., Ziqiu, Y., Xuan, D., Lai, T.-H., Jia, W.: Deploying four-connectivity and full-coverage wireless sensor networks. In: INFOCOM 2008. The 27th Conference on Computer Communications. IEEE, pp. 296–300. IEEE (2008)
45. Xu, K., Takahara, G., Hassanein, H.: On the robustness of grid-based deployment in wireless sensor networks. In: Proceedings of the 2006 International Conference on Wireless Communications and Mobile Computing, pp. 1183–1188. ACM (2006)
46. Bettstetter, C.: On the connectivity of wireless multihop networks with homogeneous and inhomogeneous range assignment. In: Proceedings of IEEE Vehicular Technology Conference (VTC), vol. 3, pp. 1706–1710 (2002)
47. Bettstetter, C.: On the minimum node degree and connectivity of a wireless multihop network. In: MobiHoc 02: Proceedings of the 3rd ACM International Symposium on Mobile Ad Hoc Networking and Computing, pp. 80–91. ACM Press (2002)
48. Wang, X., Xing, G., Zhang, Y., Lu, C., Pless, R., Gill, C.: Integrated coverage and connectivity configuration in wireless sensor networks. In: Proceedings of the 1st International Conference on Embedded Networked Sensor Systems, pp. 28–39. ACM (2003)
49. Zhang, H., Hou, J.C.: Maintaining sensing coverage and connectivity in large sensor networks. Ad Hoc Sensor Wireless Netw. **1**(1–2), 89–124 (2005)
50. Penrose, M.D.: On k-connectivity for a geometric random graph. Random Struct. Algorithms **15**(2), 145–164 (1999)
51. Bettstetter, C.: On the minimum node degree and connectivity of a wireless multihop network. In: Proceedings of the 3rd ACM International Symposium on Mobile Ad Hoc Networking and Computing, pp. 80–91. ACM (2002)
52. Gupta, P., Kumar, P.R.: Critical power for asymptotic connectivity in wireless networks. In: Stochastic Analysis, Control, Optimization and Applications, pp. 547–566. Birkhäuser, Boston (1998)

53. Wang, Y., Wang, X., Agrawal, D.P., Minai, A.A.: Impact of heterogeneity on coverage and broadcast reachability in wireless sensor networks. In: Proceedings of the 15th International Conference on Computer Communications and Networks (ICCCN), pp. 63–67. IEEE (2006)
54. Cardei, M., Wu, J.: Coverage in wireless sensor networks. Handb. Sensor Netw. 422–433 (2004)
55. Li, X.Y., Wan, P.J., Frieder, O.: Coverage in wireless ad hoc sensor networks. IEEE Trans. Comput. **52**(6), 753–763 (2003)
56. Huang, C.F., Tseng, Y.C.: The coverage problem in a wireless sensor network. Mobile Netw. Appl. **10**(4), 519–528 (2005)
57. Adlakha, S., Srivastava, M.: Critical density thresholds for coverage in wireless sensor networks. In: Wireless Communications and Networking (WCNC 2003), vol. 3, pp. 1615–1620. IEEE (2003)
58. Liu, B., Towsley, D.: A study of the coverage of large-scale sensor networks. Mobile Ad-hoc and Sensor Systems, In: Proceedings of the IEEE International Conference (2004)
59. Wang, B.: Coverage problems in sensor networks: A survey. ACM Comput. Surv. (CSUR) **43** (4), 32 (2011)

Chapter 3
Development of Home Automation System by Using ZigbeX and Atmega128 for Wireless Sensor Networks

Nik Khadijah Nik Aznan and Yeon-Mo Yang

Abstract Home automation systems have become one of the interested areas among the researchers for many years due to various advantages it offers to human as well as to the environment. Despite this fact, only relatively small number of the system implementation is available as a tutorial for undergraduate study. This paper presents a framework and a test bed of home automation systems by implementing the cost effective ZigbeX and Atmega128 with TinyOS. Through a house model, the proposed systems are able to control the lights and curtain depends on the light intensity measured by the photodiode on the ZigbeX. The design offers a user-friendly control of the system, and the system is flexible and expandable in terms of adding more sensors for future work.

Keywords Wireless sensor networks (WSNs) · Personal area networks (PANs) · TinyOS · IEEE 802.14.4 · TOSSIM · Packet analyzer · Mica-Z · Home automation

3.1 Introduction

As day by day, the earth becomes warmer and warmer, we need some solutions to improve the earth's condition. A good way that everybody can contribute is to reduce the used in the power at least at their own house. But it may be not easy for everyone to keep monitoring their electrical usage especially disabled people or elderly or even for people who left their house occasionally [1], [2].

For that reasons home automation or smart home is being developed for years. Home automation is the implementation of automatic control or computer cen-

N.K.N. Aznan · Y.-M. Yang (✉)
School of Electronic Engineering, Kumoh National Institute of Technology, Gumi, South Korea
e-mail: yangym@vivaldi.kumoh.ac.kr; ymyang@ncsu.edu

N.K.N. Aznan
e-mail: ojah87@gmail.com

S. Patnaik et al. (eds.), *Recent Development in Wireless Sensor and Ad-hoc Networks*, Signals and Communication Technology, DOI 10.1007/978-81-322-2129-6_3

tralizes control to the electrical appliances or electronic system in a building. Every household basically has lights but sometimes the lights do not need to turn on as the intensity is sufficient.

The studies on Home automation are increased rapidly in recent years. In 2006, a system for home power management [1] is proposed. The system consists of bluetooth power-controlled outlet module (BPCOM), the home server, the remote control, and the GSM channel. BPCOM is used as the network communication between master and slave, and it integrates the multiply AC power sockets and a simple plug-in low-power microcontroller to performs the switch on or off of the sockets. The user is able to control home appliances power on or off through the Internet and the GSM channel-linked to the home server in order to receive the Short Message Service (SMS).

Other than remote control, the voice recognition-based has been introduced [2]. The system will respond to voice command and control the on and off status of electrical devices. The system mainly targeted people with special needs like the elderly and the disabled. The voice that been captured by the microphone will be processed and transmitted to the voice recognition application. Then, control characters are sent to the specified electrical appliances by wireless network.

The research on home electrical appliance control application has been done by using ZigbeX on 2011 [3]. It used the Hanback Electronics ZigbeX Ubiquitous Sensor Network test bed [4, 5]. This research observed two LEDs state when GUI gives command. New programming code to control the hardware is produced by using the Tiny OS and nesC programming [6–8].

Based on the previous studies, this paper presents the development of home automation system by using ZigbeX and Atmega128. The photodiode sensors are implemented in the ZigbeX mote to monitor the light energy in a house. The LabVIEW is used to send commands to the LED and stepmotor to either turn on the light and open the curtain or turn off the light and close the curtain depends on the light intensity [6], [4], [5].

The objectives of this project is to help disable people or elderly to manage their house conveniently, to make it easy for the guardian to control the electrical appliances state while leaves kids at home and to easily handle house even when we are out of the house.

This paper is organized as follows: Section 3.2 introduces the system components including the hardware and software used. Sect. 3.3 presents the full system implementation, and this paper is concluded in Sect. 3.4.

3.2 System Modeling and Configuration

A. Wireless Sensor Networks (**WSNs**)

Wireless sensor networks (WSNs) consist of autonomous sensor to monitor and gather the physical or environment changes information like temperature, light, pressure, sound, and so on. Then, the data are transmitted through network to the corresponding location [4, 5].

B. ZigbeX

ZigbeX is using the Zigbee standard [9], which can be used for wireless sensor network principles. The ZigBee is a radio frequency standard based on IEEE 802.15.4 where specified the physical layer and MAC layer for low-rate wireless personal area network (LR-WPAN). Zigbee suitable for wireless control such as building automation, personal health care, industrial control, consumer electronics, and so on as it has high-level communication protocols but using small- and low-power digital radios. It supports simple devices with minimum energy consumption.

Table 3.1 shows comparison between Zigbee standard and other wireless standard. The low data rate of Zigbee compared to others make it suitable for the controllers and sensors application. Low complexity with high range are the caused that make Zigbee always be used for automation industry.

ZigbeX is a platform for the ubiquitous computer. It can be used with Atmega 128, and it supports Tiny OS. It provides antenna port for distance communication and extended ports for sensors.

C. Tiny OS

Tiny OS is an open-source operating system (OS), which can be used in the wireless sensor network (WSN) [10]. It is an event-driven operating system with limited resources to reduce the energy consumption for sensor network nodes. It is in the sleep mode all the time and will turn on automatically whenever there is a process. So, the energy used is very small.

Tiny OS is used as nesC programming language for the application which is the extension of C. nesC consists of 2 components which are modules and configurations. nesC separate the construction and the composition of the programs. The construction is made out of components and are assembled to form whole programs. Moreover, it is designed for better code generation and analysis, so it is expected to generate the code by the whole-program compilers. Figure 3.1 shows the structure of the nesC.

Table 3.1 Comparison between Zigbee standard and other wireless standard

Parameters	Zigbee	802.11	Bluetooth
Data rate	20, 40, 250 Kbps	11, 54 Mbps	1 Mbps
Range (meter)	10–100	50–100	10
Network topology	Ad hoc, peer to peer, star or mesh	Point to hub	Ad hoc, very small networks
Operating frequency	2.4 GHz	2.4 and 5 GHz	2.4 GHz
Complexity	Low	High	High
Power consumption	Very low	High	Medium

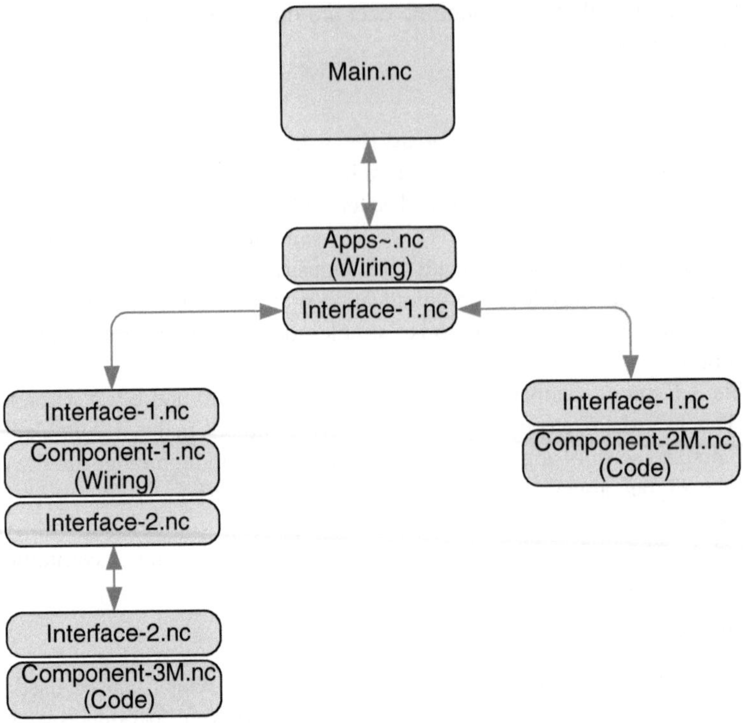

Fig. 3.1 The structure of nesC

D. LabVIEW

LabVIEW is software for system design platform that provides tools for visual programming language to create and deploy measurement and control systems through hardware integration. It used G code, which is a dataflow programming language. The program will execute the graphical block diagram and flowchart. LabVIEW includes a lot of drivers and abstraction for different types of instrument and busses.

E. Atmega128

Atmega128 is a microcontroller based on AVR RISC architecture. It is high performance, low-power CMOS, and an 8-bit chip microcontroller. AVR was developed by Atmel and by one of the first microcontroller for program storage that use on-chip flash memory. It has 6 groups which are tinyAVR, megaAVR, XMEGA, Application-specific AVR, AVR with FPGA, and 32-bit AVR. Atmega128 is in the megaAVR groups with 4–512-kB program memory and 28–100 pin package. As a microcontroller, Atmega128 is used to control the action of the system.

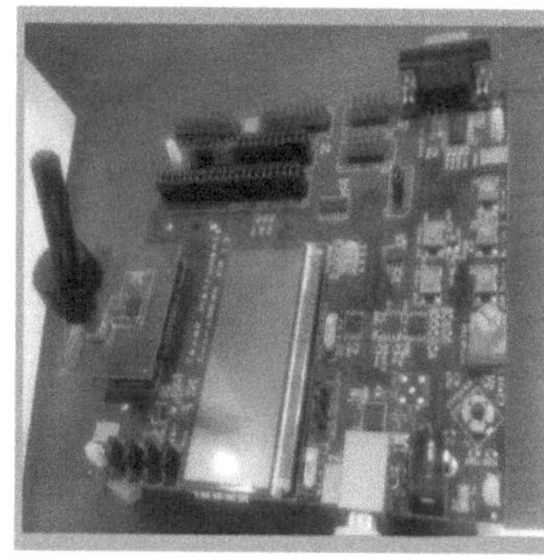

Fig. 3.2 Packet sniffer CC2520EM [11]

F. Packet Sniffer

Packet sniffer is used to capture RF packets with a listening RF device. It can be used to detect network problem and confirm the data transmission. Texas Instrument's packet sniffer is used to analyze the Zigbee communication. (Figure 3.2) (Fig. 3.3)

Figure 3.4 shows the successful transmission where the FCS is OK, while Fig. 3.5 shows the unsuccessful transmission where the FCS is error. FCS is the frame check sequence, which is a frame for error detection.

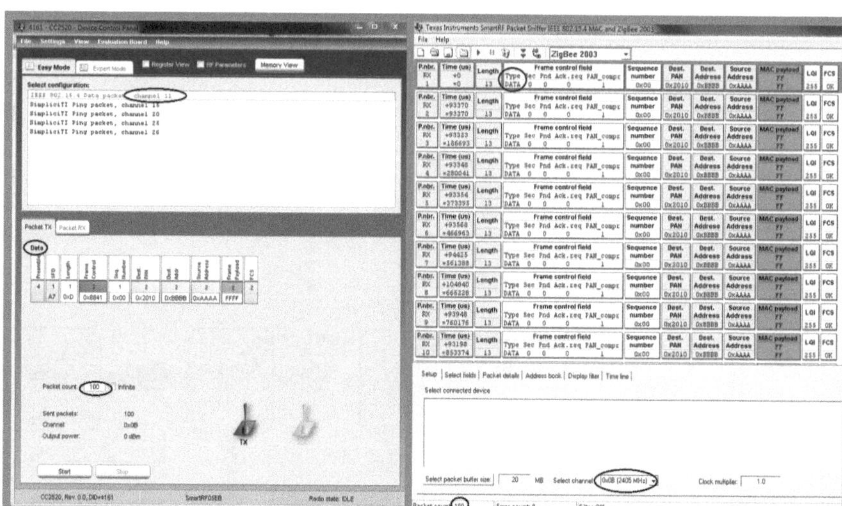

Fig. 3.3 Comparison between packet transmit and packet sniffer

P.nbr.	Time (us)	Length	Frame control field					Sequence number	Dest. PAN	Dest. Address	Source Address	MAC payload	LQI	FCS
RX	+0		Type	Sec	Pnd	Ack.req	PAN_compr					FF		
1	=0	13	DATA	0	0	0	1	0x00	0x2010	0xBBBB	0xAAAA	FF	255	OK

Fig. 3.4 Successful transmission

P.nbr.	Time (us)	Length	Frame control field					Sequence number	Dest. PAN	Source Address	LQI	FCS
RX	+186549675		Type	Sec	Pnd	Ack.req	PAN_compr					
6	=1670667716	46	R100	0	1	1	1	0x92	0x053D	0x143B	68	ERR

Fig. 3.5 Unsuccessful transmission

3.3 System Implementation and Experimental Results

This section explains the results of system implementation and the experimental results under WSNs.

Figure 3.6 explains the system implementation, which consist of three stages; data collection, observation or control, and action taken. The sensor in ZigbeX mote will monitor the changes of the environment and collect the data. The sensor used for this project is photodiode. The data will be sent through wireless to the ZigbeX mote that is connected to a PC by using serial UART.

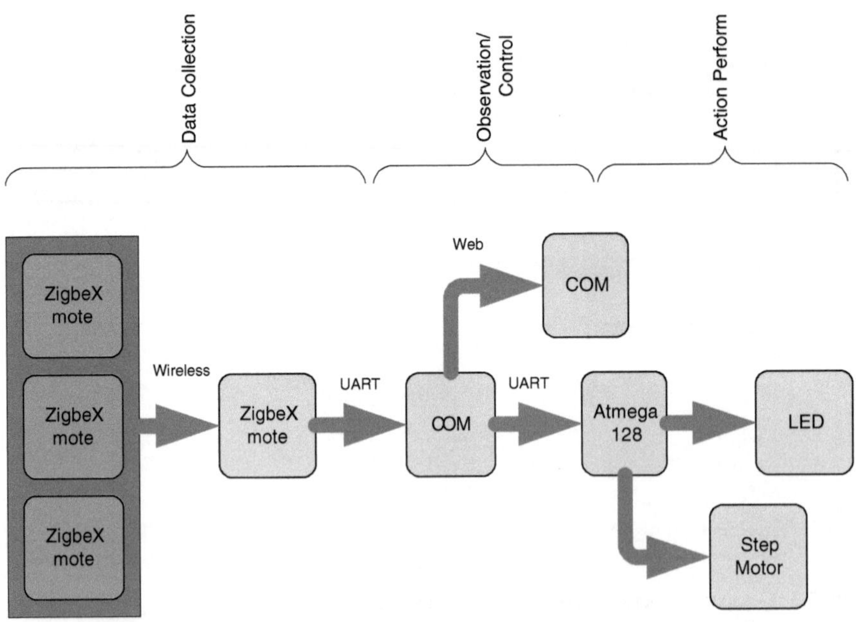

Fig. 3.6 Overview block diagram of the full system implementation

The ZigbeX used Tiny OS as the operating system. Tiny OS communicates with computer by using serial UART. The data received in a computer will be read by LabVIEW that will program commands for the microcontroller. If the reading of photodiode gives the high values which it has high intensity, the LED will be turned off and the curtain will be closed. While, when the photodiode has low intensity, the LED will be on and the curtain will be open.

To apply this implementation, a model house (Fig. 3.7) is built. The LEDs are used as lights, while stepmotors are used as the automatic curtain. The photodiode sensors on ZigbeX motes are placed at some spots of the house. The LEDs and stepmotors are connected to the microcontroller Atmega128 to get the command to perform the action.

By using software like in Fig. 3.8, the data transmission from the ZigbeX mote can be viewed. At the small box, it is the address node 01, 02, and 03 of the sensors. The box beside, in the third line, the sensors are shown. BA 02 B9 02 is for illuminance, 17 00 17 00 is for infrared, 1B 00 1B 00 is for temperature, and 18 00 18 00 is for humidity.

In order to be a user-friendly system, a user action screen interface (Fig. 3.9) is designed so that user can choose either to let the system works automatically or manually. The user can switch the LED on or off by using the interface. Other than that user also can adjust the brightness of the LED.

Fig. 3.7 The house model for the implementation

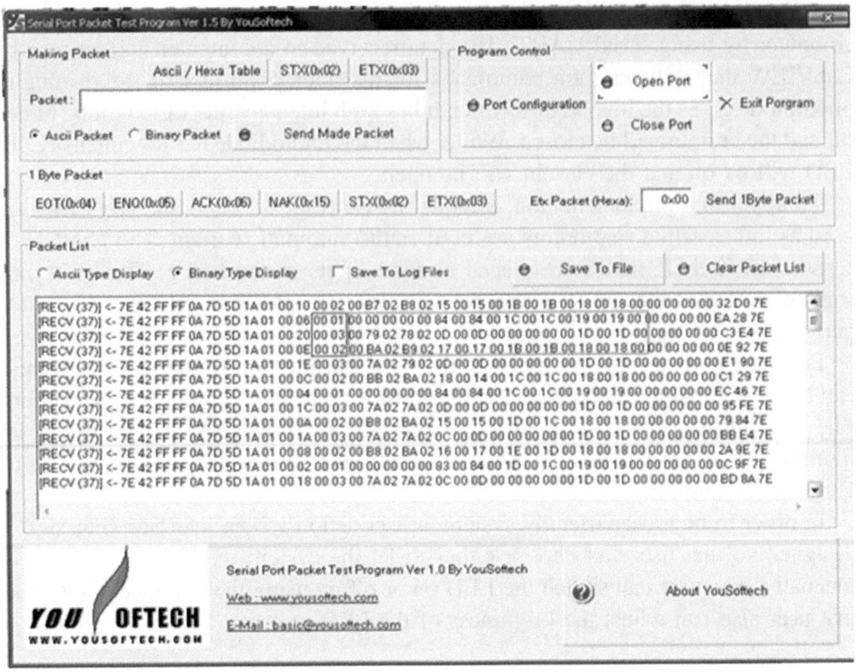

Fig. 3.8 Serial port packet transmission

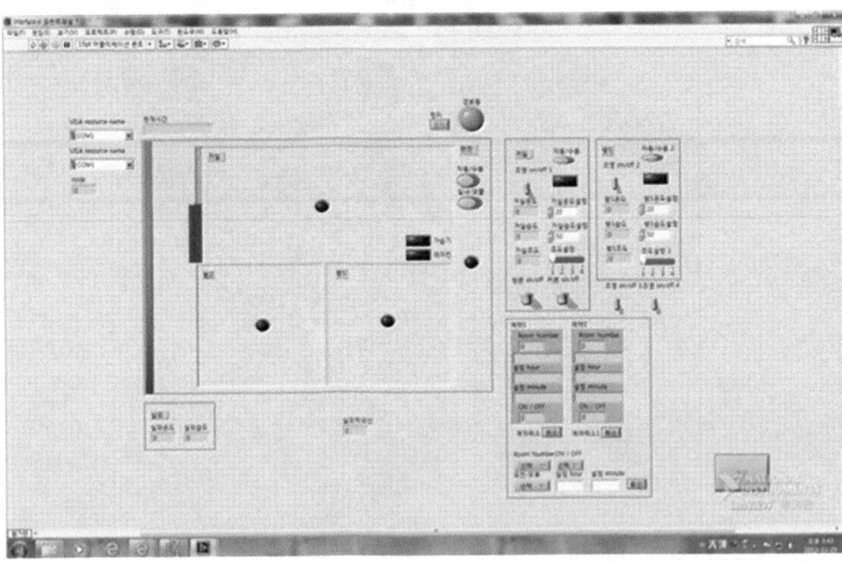

Fig. 3.9 User action screen interface

3.4 Conclusion

In summary, we implemented the ZigbeX and Atmega128 in designed the Home Automation system. The ZigbeX used Tiny OS as operating system and nesC to program it, while Atmega128 used LabVIEW as a design platform. When the light intensity in the house is high, the light will be turned off and the curtain will be closed automatically. When the light intensity in the house is low, the light will be turned on and the curtain will be opened automatically. With this system, it eases people to manage their house, and at the same time, it reduces the power consumption especially people with special needs. This system can be improved by considering the light intensity from outdoor as well as by using the natural source which is sunlight; we may be able to reduce even more power consumption. This system can be developed more for other appliances like fan, air-conditioner, heater, humidifier, and others. The implementation of this system also can be used in other places like office, school, and even for the outdoor environment.

Acknowledgments This paper was supported by Research Fund, Kumoh National Institute of Technology (www.kumoh.ac.kr).

References

1. Lin, C.H., Bai, Y.W., Liu, M.F., Lin, M.B.: Remotely controllable outlet system for home power management.In: IEEE 10th International Symposium on Consumer Electronics, ISCE '06, pp. 1–6 (2006)
2. Kasim, M., Isa, N.M.I., Daud, J.R.: Home electrical appliance control application using Zigbex Ubiquitous Sensor Network. In: IEEE International Conference on System Engineering and Technology (ICSET), pp. 143–147 (2011)
3. AlShu'eili, H., Gupta, G.S., Mukhopadhyay, S.: Voice recognition based wireless home automation system. In: 4th International Conference on Mechatronics (ICOM), pp. 1–6 (2011)
4. Yang, Y.-M., Jeon, I.: An efficient self routing scheme by using parent-child association for WSNs, LNCS 5370, pp. 785–794. Springer, Berlin (2008)
5. Yang, Y.-M., Kim, S.: An Improved self-routing scheme by using Parent-child relationship and beacon-olny period for WSNs. Inderscience IJICT **2**(4), 279–301 (2010)
6. Naruephiphat, W., Charnsripinyo, C., Satienpaisarn, S., PromYa, R.: Applying wireless sensor network for power consumption monitoring. In: 9th International Conference on Electrical Engineering/Electronics, Computer, Telecommunications and Information Technology (ECTI-CON), pp. 1–4 (2012)
7. Gill, K., Yang, S.H., Yao, F., Liu, X.: A ZigBee-Based Home Automation System. IEEE Trans. Consum. Electron. **55**, 422–430 (2009)
8. Felix, C., Raglend, I.J.: Home automation using GSM. In: International Conference on Signal Processing, Communication, Computing and Networking Technologies (ICSCCN), pp. 15–19 (2011)
9. Zigbee Alliance.: Zigbee. http://www.zigbee.org/. Retrieved 3 Jan 2013
10. Wireless embedded system.: TinyO. http://webs.cs.berkeley.edu/tos/download.html, University of California, Berkeley (2005). Retrieved 17 Dec 2012

11. Texas Instruments.: SmartRF Protocol Packet Sniffer. http://www.ti.com/tool/packet-sniffer. Retrieved 7 Jan 2013
12. Hanback Electronis Co., LTD.: HBE- Zigbex. http://www.hanback.cn/en/index.asp. Retrieved 3 Jan 2013
13. IEEE 802, May 10, 2013, IEEE 802.15 WPAN Task Group 4. http://www.ieee802.org/15/pub/TG4.html. Retrieved 31 Jan 2013

Chapter 4
Efficient Coordination and Routing Protocol for Wireless Sensor and Actor Networks

Biswa Mohan Acharya and S.V. Rao

Abstract Wireless sensor and actor networks (WSAN) are composed of low-powered, low-resources, densely deployed sensor nodes and high-powered, resource-rich, sparsely deployed actor nodes to perform distributed sensing and acting tasks, respectively. Sensor nodes sense the physical phenomenon and report to actor nodes at earliest possible time for appropriate action by the actor nodes. The major objective of WSAN is to have the desired action correspondent to the reported event with higher precession. Due to the coexistence of sensor and actor nodes, one of the major challenges in such networks is communication and coordination mechanisms among the sensor and actor nodes. We have considered the problem of communication and coordination and proposed an efficient model based on geometric structure called Voronoi diagram. Our protocol is based on clustering (virtual grid) and Voronoi region concept. Simulation results demonstrate that the proposed protocol outperforms in terms of throughput, packet delivery ratio, average delay, and normalized routing overhead.

Keywords Wireless sensor and actor networks · Voronoi region · Coordination protocol

4.1 Introduction

Wireless sensor network (WSN) is a collection of large number of tiny sensor nodes communicating together to achieve an assigned task. A sensor node is a tiny battery-operated device capable of sensing, communicating, and computing. Sensor nodes

B.M. Acharya (✉)
Department of Computer Applications, Institute of Technical Education and Research, Siksha 'O' Anusandhan University, Khandagiri Square, Bhubaneswar, India
e-mail: biswaacharya@soauniversity.ac.in

S.V. Rao
Department of CSE, IIT Guwahati, Guwahati, India
e-mail: svrao@iitg.ac.in

S. Patnaik et al. (eds.), *Recent Development in Wireless Sensor and Ad-hoc Networks*, Signals and Communication Technology, DOI 10.1007/978-81-322-2129-6_4

are responsible for detecting events and transmitting the same to sink node. The sink node is usually not reactive in nature and presents at a far distance from the event area. However, for some time, in critical applications such as forest fire control and military surveillance applications, upon receiving the information from sensor nodes, there is a need of early action to deal with the situation. In such cases, we need powerful nodes having actuation units (mechanical units) attached with them, deployed along with the sensor nodes, which can receive information from sensor nodes and take actions accordingly. These networks are an extension of traditional WSNs and known as wireless sensor and actor networks (WSAN). WSAN are capable of observing the physical environment, processing the data, making decisions based on the observations, and performing appropriate actions and therefore have recently attracted much attention of researchers due to its wide ranges of applications.

WSAN refer to a group of sensor and actor nodes linked by wireless medium to perform distributed sensing and acting tasks. In WSAN, the phenomena of sensing and acting are performed by sensor and actor nodes, respectively. Sensor nodes gather information about the physical environment by automatically collaborating to sense, collect, and process data, and transmit those sensed/processed data to some specific actor nodes. Actor nodes take decisions and then perform appropriate actions upon the environment. Sensor nodes are low-cost, low-powered devices with limited sensing, computation, and wireless communication capabilities, whereas actor nodes are resource-rich nodes equipped with better processing capabilities, higher transmission powers, and longer battery life. Also, actor nodes are physically equipped with actuation units (mechanical device) in order to take action in the event area. The density of sensor nodes deployed uniformly in a target area may be in the order of hundreds or thousands to perform collaborative sensing and communicating task, but such a dense deployment is usually not necessary for actor nodes, because actor nodes have higher capabilities and can act on large areas.

The objective of WSAN is to get the event information from the physical environment where they are deployed, process the sensed data, take decision based on those data, and perform the required action to deal with the detected event effectively. When an event is detected, a sensor node signals this event information to an actor node to deal with. For example, in case of forest fire, sensor nodes relay the exact location of fire to the actor nodes so that actors having water sprinkler devices can sprinkle water and extinguish the fire. Due to the coexistence of heterogeneous nodes, sensors, and actors, there are several challenges in WSAN. One of the major challenges for efficient communication is the coordination among sensor and actor nodes. For example, due to dense deployment of sensor nodes, an event is detected by many sensor nodes and all of them communicate to actor nodes which may lead to more number of transmissions, congestion in the network, and redundant action by actor nodes. Hence, WSAN need an efficient coordination among sensor and actor nodes. That is sensor–sensor, sensor–actor, and actor–actor coordination.

The main goal of sensor–sensor coordination is to collect information in the event area. In order to maximize the lifetime of sensor nodes, there is the need of following the sleep and wake-up schedule among sensor nodes in such a way that at any instant of time, minimum number of sensor nodes remain awake which covers the entire region for sensing and communication purpose. The challenge here is that how to choose the optimal number of sensor nodes to achieve this objective. Hence, we realized that clustering is required to minimize energy consumption and maximize network lifetime.

The most important characteristic of sensor–actor communication is to provide low communication delay due to the proximity of sensor and actor nodes. One of the main requirements of sensor–actor communication is to consume low energy. In some applications such as in fire, the communication traffic is typically delay sensitive. Therefore, another main requirement of sensor–actor communication is to support real-time traffic. An additional requirement for communication in WSAN is the need to ensure ordering of event data reported to the actors. For example, if there are two sensor nodes reporting two different events to an actor or some actors in overlapping regions, then the reporting of those events must be done in sequence in which the events were detected so that the correctness of the actions on the environment is guaranteed. This is known as the ordered delivery of information collected by the sensors. Another important consideration is that if there are multiple sensors reporting an event, then the information from different sensor nodes may arrive at the concerned actor nodes approximately at the same time. This may be necessary to ensure that the action is performed once and in the entire event region.

Sensor–sensor coordination is for ensuring that only optimal number of sensor nodes remains active and rest can be in sleep mode to conserve energy and also reduce number of communications. Sensor–actor coordination is for quick event reporting to appropriate actor node. Actor–actor coordination is to avoid redundant action in overlapped action area and a reliable action.

Hence, communication and coordination among the sensor and actor nodes is a major design goal for WSAN.

Actor–actor communication occurs in the following situations:

- The actor node receiving sensor node data may not act on the event area due to small action range or insufficient energy.
- One actor node may not be enough to perform the required action; thus, other nearby actor nodes should be triggered.
- If multiple actor nodes receive the same event information and there is an action threshold, these actor nodes should talk to each other in order to decide which one of them should perform the action.
- In case of multiple events occurring simultaneously, task assignment can be done via actor–actor communication. Also, it may be desired that the tasks are executed sequentially. This constraint is referred to as ordered execution of tasks.
- After an actor node receives event information if the event is spreading to other actor node's acting area, the actor node can transmit the sensor data or action command to those actors. In this way, there will be no need for sensor nodes in those areas to transmit event information to the nearby actor nodes as they will be forwarded by initial set of actor nodes. This is an alternative to the tracking problem where the actor nodes handle different locations of the events. All of

the above situations which indicate the necessity of actor–actor coordination converge to the issue of deciding which actor(s) should execute which action(s). Solution to this can be given by exploiting the coordination between actor nodes. Actor nodes should, whenever possible, coordinate strongly with each other in order to maximize their overall task performance.

We address these challenges by proposing a model for WSAN which organizes the WSAN in Voronoi regions such that each region contains only one actor node and it is the nearest to all sensor nodes inside that region. We also reduce the number of communication among the sensor nodes by assuming grid architecture and following sleep/wake-up schedule among sensor nodes inside a grid which can increase the network life.

4.2 Literature Review

A comprehensive survey on various research issues of WSAN and the future aspects has been pointed out by the authors in [1]. The authors have discussed the architecture of WSAN and the different issues such as sensor–sensor, sensor–actor, and actor–actor coordination along with the research challenges. The unique characteristics of WSAN such as real-time requirements and coordination issues, which is our main focus, are also discussed at length.

A coordination protocol framework for WSAN is addressed with a proposed sensor–actor coordination model based on event-driven clustering paradigm where cluster formation is triggered by an event so that clusters are created on the fly to optimally react to the event and provide reliability with minimum energy consumption [2].

In [3], coordination and communication problems with mobile actors are studied and a hybrid location management scheme is introduced to handle mobility of actors with minimum energy consumption. Actor nodes broadcast location updates based on Voronoi region as their scope of communication.

Authors in [4] focus on three aspects of coordination namely sensor–sensor coordination, sensor–actor coordination, and actor–actor coordination. As the numbers of nodes in WSAN are quite large, researchers show that sensor network protocols and algorithms should be scalable and localized.

An energy-efficient layered routing scheme is described in [5] for semi-automated architecture, where the network field is divided into different-sized (overlapped) actor fields, which covers all the sensor nodes. The communication is transformed into two layers. In each actor field, actor is the center of the network and sensor nodes transmit information to the actor node. Actor nodes communicate with other actor nodes and sink directly, but sensor nodes communicate with other sensor and actor nodes hop-by-hop.

A sensor–sensor coordination protocol for WSAN based on clustering and Voronoi region concept is proposed in Bouhafs et al. [6]. This protocol creates clusters consisting of sensors detecting the same event and forwards to the nearest actor. However, the event-driven clustering approach maintains many paths between cluster members and cluster heads, which is possible because of more number of communication. Also here, there is periodic transmission of information which maintains the path among the sensors.

4.3 The System Model

Our protocol considers few assumptions for simplicity. Both sensor and actor nodes are assumed to be immobile in nature. Actor nodes have much more efficiency in terms of energy level and transmission power as compared to sensor nodes. Sensor and actor nodes are deployed uniformly in the network area with high density of sensor nodes and low density of actor nodes. Every sensor node associates itself to one of the actor nodes to which it is nearest, which leads to the construction of Voronoi regions around actor nodes. It is also assumed that all the nodes (both sensor and actor) are equipped with GPS-enabled devices, and hence, each node is aware of its own location information.

For efficient energy utilization, we need to keep optimal number of sensor nodes awake which covers the entire network area. Keeping all sensor nodes active makes the network reliable, but it consumes lot of energy, and also, the same event is detected by many sensor nodes. This leads to other problems such as reduced network lifetime and congestion. In order to conserve energy and increase network life, we schedule nodes such that there is only one node active in each grid. The protocol achieves this by dividing the network area into fixed-size square grids, and one sensor is chosen in each grid which remains active at a time. Every sensor node is at least sufficient enough to sense a grid area to which it belongs, and also, every sensor node is capable enough to communicate directly to the farthest node in its neighboring grid. Each sensor node sends the information about events in multi-hop manner to actor, to reduce the energy consumption for communication. For complete coverage and communication, we assume the communication radius R_C (distance from X to Z in Fig. 4.1) of sensor nodes is at least twice of the sensing radius R_S (distance from X to Y in Fig. 4.1), i.e., $R_C \geq 2R_S$.

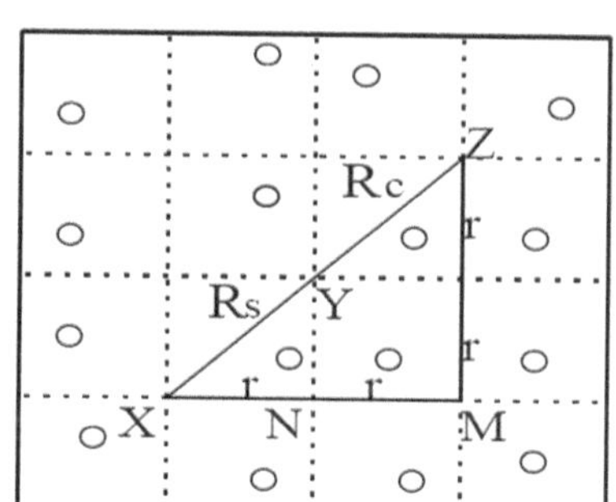

Fig. 4.1 Grid size computation

This assumption of relationship between sensing radius and communication (transmission) radius ensures the coverage and connectivity of our protocol as described in theorem 1 of [7].

The size of a grid can be calculated as follows.

Let n nodes are deployed within a topology area T and let the grid has r units (distance between X and N) on a side as shown in Fig. 4.1. The value of r can be calculated as

$$r \leq \frac{R_c}{2\sqrt{2}}$$

4.4 The Procedure

Our procedure consists of three phases, and they are as follows:

4.4.1 Initialization Phase

Sensor and actor nodes are deployed uniformly throughout the network area under consideration. Actor nodes create Voronoi region around themselves consisting of all the sensor nodes which are closer to itself than any other actor nodes as shown in Fig. 4.2.

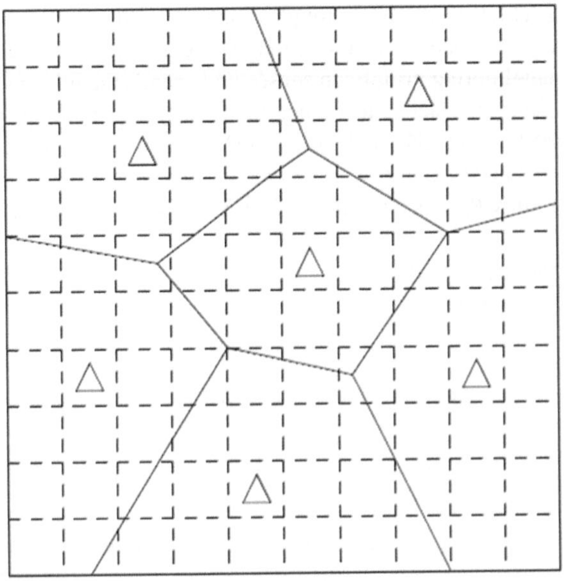

Fig. 4.2 Network area divided into grids and Voronoi regions

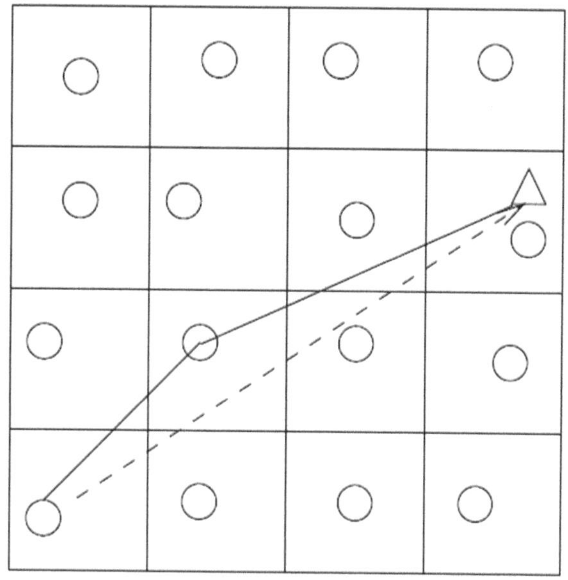

Fig. 4.3 Greedy forwarding from sensor to actor

The predefined route from sensor node to actor node reduces the cost of route establishment during occurrence of events or forwarding of event information. Every sensor node gathers the information regarding its neighboring sensor nodes inside the same grid and associated with the same actor node for node scheduling, which is a major requirement for sensor–sensor coordination.

4.4.2 Detection and Reporting Phase

When an event occurs, it is detected by the active sensor node inside the grid(s) in which it occurs and transmitted along the optimal path toward the nearest actor node as shown in Fig. 4.3. This phase refers to sensor–actor coordination in WSAN. The active sensor nodes in other grids which are one hop away from actor node directly transmit without transmitting to the active sensor node inside actor node's own grid.

4.4.3 Action Phase

Actor node aggregates and processes to take the required action. It is the responsibility of actor node to ensure that all the locations are covered. When new event information is received from within its Voronoi region by the actor node after initiating action, it adjusts its power level to take care of the new locations. The objective of our protocol is to send the sensed information at the earliest to the actor node with reduced redundancy, and for that, we propose the data aggregation approach.

4.5 Implementation and Result Analysis

To measure our success in meeting the design goals, we simulated our approach with NS-2.33 and analyzed the performance in terms of throughput, packet delivery ratio, average delay, and normalized routing overhead.

We observed the above-mentioned performance metrics with respect to the variation of traffic load (CBR packets) in the network and also with respect to the packet size for three different number of actor nodes.

Figure 4.4 shows the relationship between CBR packet interval (seconds) and average delay (seconds) for three cases of number of actors. Here, when we increase the number of actor nodes, the average delay decreases, because with more number of actor nodes, the actor nodes becomes closer to the event area, and hence, delay decreases. Based on the type of application and delay bound, we need to figure out the number of actor nodes, because the cost of actor nodes is high. For each actor case, when we increase the load after a certain value (0.5 in this case), there is a sudden rise in the average delay value because of collision.

Figure 4.5 describes the relationship between traffic load and throughput (bps) of the network. It clearly shows that the more is the number of packets sent, the more is the throughput. With change in number of actors, there is not that much difference for CBR interval of up to 0.5 from 2.0. But when the CBR interval is 0.25 or more, there is a remarkable change which depicts our limitation of load.

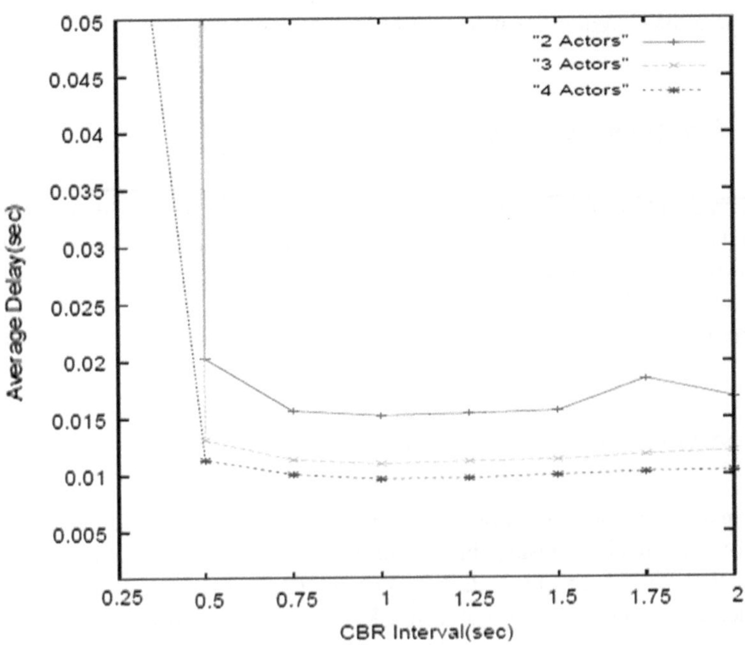

Fig. 4.4 CBR versus average delay

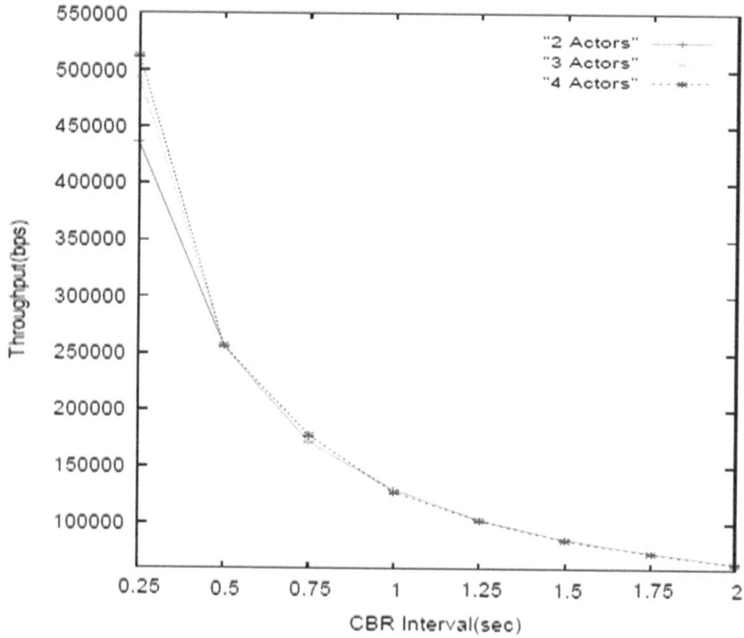

Fig. 4.5 CBR versus throughput

Figure 4.6 shows the relationship between CBR packet interval (seconds) and the packet delivery ratio (percentage). From the figure, it is quite clear that our assumption of keeping one sensor node active per grid keeps the system connected and there is very less packet drops in the network. Hence, it can be concluded that there is less data loss in our protocol up to the limitation of traffic load. However, when the traffic load is more (CBR interval of 0.25 in this case) than our limitation, there is a sudden drop in delivery ratio. Also, the figure claims that the more the number of actors, more is the delivery ratio because the actors become closer to the event area, and hence, packet drop rate is reduced.

Figure 4.7 represents the number of routing packets sent over number of data packets. It is the routing overhead which counts for the cost of the network. From the figure, it can be concluded that when load increases, the overhead decreases because for maintaining the path and network setup, the number of control packets remain fixed, but with increase in load, the number of data packets increase and hence the decrease in overhead. However, as far as the number of actor nodes is concerned, for the optimal load, the difference in overhead is not that big. So depending on the feasibility of the requirement and cost, the number of actors can be effectively chosen.

We also simulate the protocol with different size of CBR packets. We vary the CBR packet size from 64 bytes to 512 bytes. Different graphs of all the performance

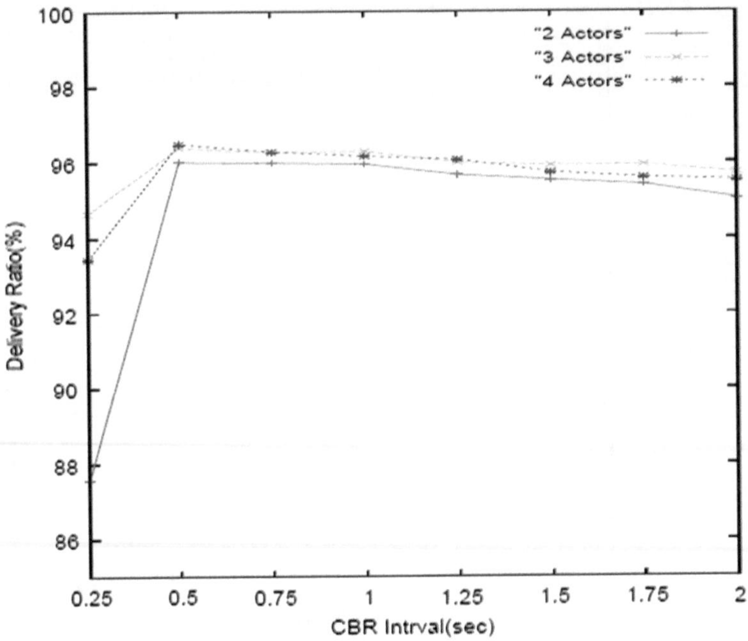

Fig. 4.6 CBR interval versus packet delivery ratio

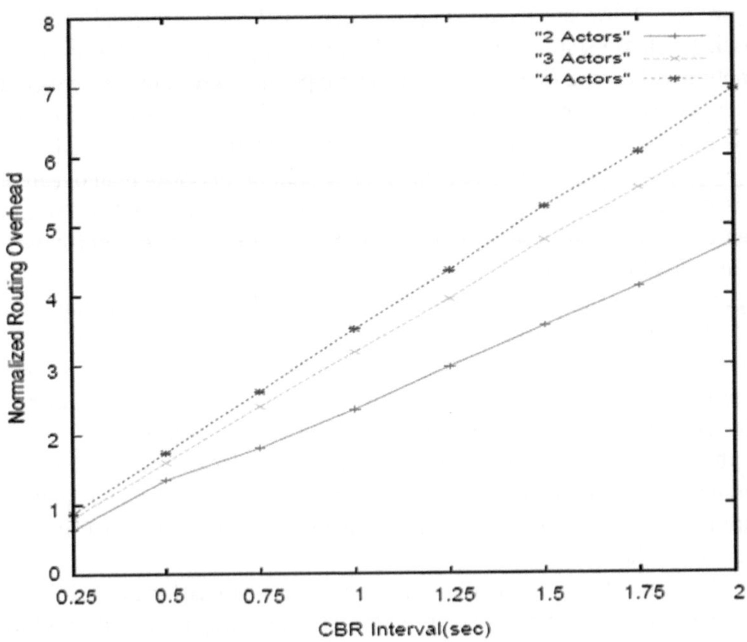

Fig. 4.7 CBR interval versus normalized routing overhead

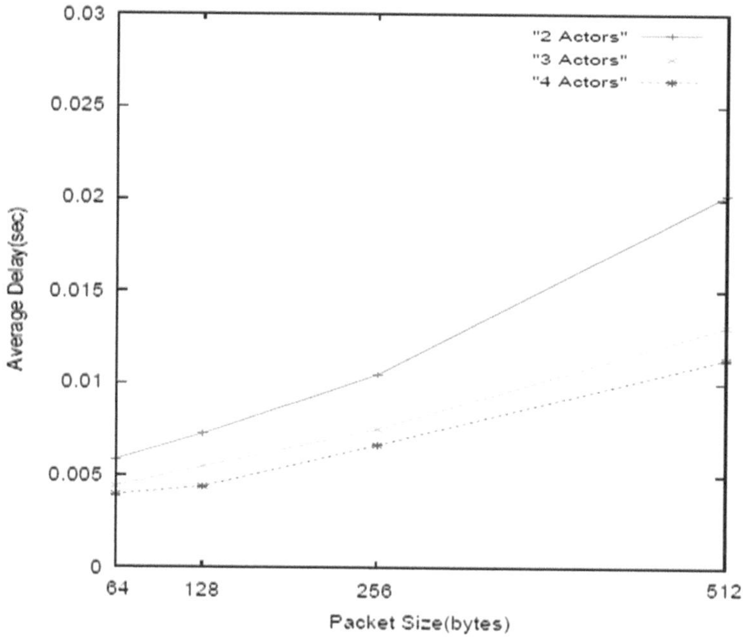

Fig. 4.8 Packet size versus average delay

metrics (as described in the above section) against payload clearly indicates that the protocol performs as expected.

Figure 4.8 shows the relationship between CBR packet size and average delay for all the cases. So it is an indication that with increase in number of actor nodes for any packet size, the delay is reduced, because actor nodes become closer to event area.

Figure 4.9 depicts the relationship between CBR packet size and the packet delivery ratio. From this figure, also, it is clear that there is very less number of packet loss in the network with increase with number of actor nodes. The graph shows that the delivery ratio is hardly affected with change in packet size.

Figure 4.10 represents the normalized routing overhead against the packet size. As discussed earlier, with increase in number of actor nodes, there is an increase of number of Voronoi regions and hence increase in number of control packets. However, there is almost no change in overhead when we increase the packet size which is quite obvious because it matters with number of packets rather than the size of packets.

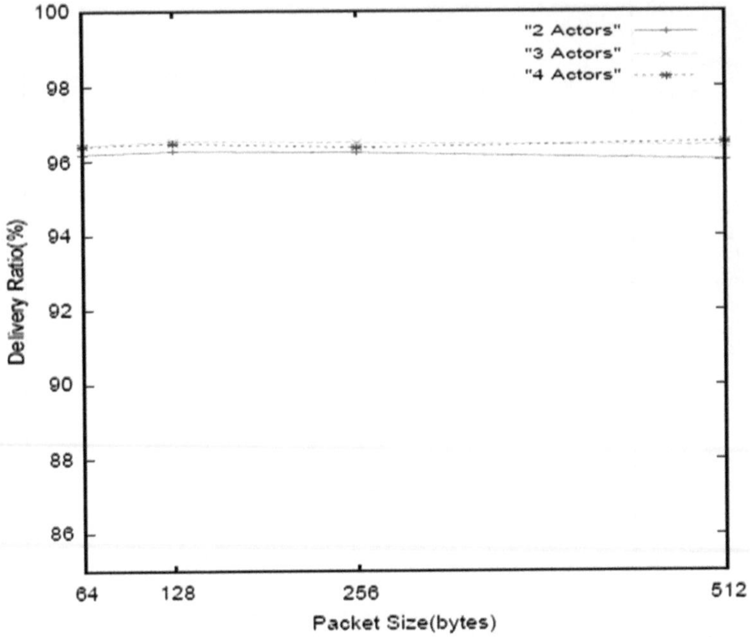

Fig. 4.9 Packet size versus packet delivery ratio

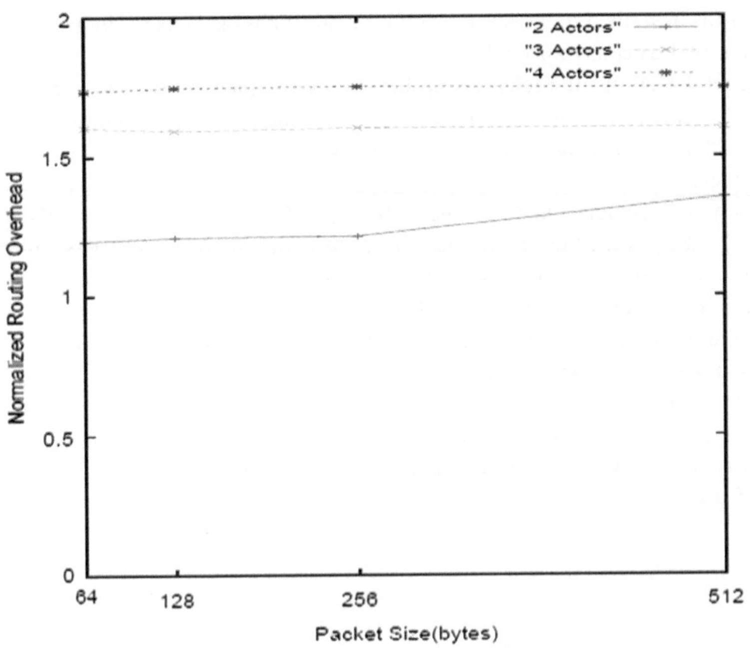

Fig. 4.10 Packet size versus normalized routing overhead

4.6 Conclusion

Our approach tries to improve the throughput and packet delivery ratio, while minimizing routing overhead and delay, by choosing a stable path for packet forwarding through static routes. The protocol emphasizes on coordination among sensor and actor nodes for WSAN based on Voronoi region and virtual grid and reduce congestion and redundancy in the network. The protocol divides the network area into grids, and in each grid, only one sensor node remains active. Sensor nodes gather event information and send it to nearest actor node based on Voronoi region. The important characteristics such as real-time requirements and efficient utilization of available node energy are also taken into consideration.

The advantage is that as for each sensor node there is only one closest actor node based on the Voronoi region, there will be no problem of deciding the actor node to which data need to be transmitted upon detecting an event which is a major challenge in WSAN. WSAN need synchronization mechanisms between sensor nodes that detect the same event in order to avoid more than one actor node invocation and overlapping between them. For that reason, sensor nodes need to share and coordinate their information between themselves before sending data to actor node, which lead us to have virtual grid concept similar to clustering approach. Sensor–sensor coordination issue is taken care by dividing the network area into grids and scheduling nodes within each grid to keep only one sensor node active at any time for event detection and reporting. Sensor–actor coordination is achieved by data aggregation among sensors and following simple greedy forwarding approach.

References

1. Akyildiz, I.F., Kasimoglu, I.H.: Wireless sensor and actor networks: research challenges. Ad Hoc Netw. **2**, 351–367 (2004). Elsevier
2. Melodia, T., Pompili, D., Gungor, V.C., Akyildiz, I.F.: A distributed coordination framework for wireless sensor and actor networks. In: Proceedings of 6th ACM international symposium on mobile ad hoc networking and computing, May (2005)
3. Melodia, T., Pompili, D., Akyildiz, I.F.: A communication architecture for mobile wireless sensor and actor networks. In: Proceedings of IEEE SECON (2006)
4. Yuan, H., Ma, H., Liao, H.: Coordination mechanism in wireless sensor and actor networks. In: Proceedings Of the First International Multi- Symposiums on Computational Sciences (Imsccs'06) (2006)
5. Peng, H., Huafeng, W., Dilin, M., Chuanshan, G.: Elrs:an energy efficient layered routing scheme for wireless sensor and actor networks. In:Procedings of the 20th IEEE International Conference on Advanced Information Networking and Applications (AINA'06) (2006)
6. Bouhafs, F., Merabti, M., Mokhtar, H.: A coordination protocol for wireless sensor and actor networks. In: PGP Net (2006)
7. Aurenhammer, F.: Voronoi diagrams: a survey of a fundamental geometric data structure. ACM **23**, 345–405 (1991)

Chapter 5
Performance Comparison of BEMRP, MZRP, MCEDAR, ODMRP, DCMP, and FGMP to Achieve Group Communication in MANET

M. Rajeswari, P. Uma Maheswari and S. Bhuvaneshwari

Abstract In this paper, we present a comparative performance of six multicast protocols for mobile ad hoc networks—BEMRP, MZRP, MCEDAR, ODMRP, DCMP, and FGMP focusing on the effects of changes such as the increasing number of receivers or sources and increasing the number of nodes. Although some simulation results of MANET protocols have been published before, these six protocols have not been compared in isolation. In recent years, a number of new multicast protocols have been proposed for ad hoc networks. Due to the limited transmission range of wireless mobile nodes, multiple network hops may be required for one node to exchange the data with another across the network. Multicasting is an efficient method for implementing group communications. But, it is difficult to achieve this, due to the difficulty in group membership management over a dynamic topology. A systematic performance evaluation of these protocols is done by performing certain simulations under NS-2.

Keywords Multicasting · BEMRP · MZRP · MCEDAR · ODMRP · DCMP · FGMP

5.1 Introduction

An ad hoc network is a wireless network formed by wireless nodes without infrastructure. In such a network, the nodes are mobile and can communicate dynamically in an arbitrary manner. The network is characterized by the absence of

M. Rajeswari (✉) · S. Bhuvaneshwari
Angel College of Engineering & Technology, Tirupur, Tamil Nadu, India
e-mail: rajimanickam@gmail.com

S. Bhuvaneshwari
e-mail: bhuvimk28@gmail.com

P. Uma Maheswari
Info Institute of Engineering, Coimbatore, Tamil Nadu, India
e-mail: umayam2003@yahoo.co.in

S. Patnaik et al. (eds.), *Recent Development in Wireless Sensor and Ad-hoc Networks*, Signals and Communication Technology, DOI 10.1007/978-81-322-2129-6_5

central administration devices such as base stations or access points. Furthermore, nodes should be able to enter or to leave the network easily. In these networks, the nodes act as routers. They play an important role in the discovery and maintenance of the routes from the source to the destination or from a node to another one. This is the principle challenge to such a network. If link breakages occur, the network has to stay operational by building new routes. Multi-hopping is used to increase the overall network capacity and performance where one node can deliver data on behalf of another one to a determined destination. Thus, the problem of range radio is solved. A mobile ad hoc network (MANET) represents a system of wireless mobile nodes that can self-organize freely and dynamically into an arbitrary and temporary network topology. On one hand, they can be quickly deployed anywhere at any time as they eliminate the complexity of infrastructure setup.

Multicasting [1] is an important communication pattern that involves the transmission of packets to a group of two or more hosts, and thus is intended for group-oriented computing [2, 3]. Applications of MANETs include battlefield communications [4], disaster recovery, collaborative and distributed computing, emergency operations, co ordinate task scheduling (such as earth moving or construction), vehicular communication for traffic management, data and information sharing in difficult terrain, and extension of the infrastructure-based wireless networks. The following are the issues in ad hoc wireless networks.

Medium Access Scheme The primary responsibility of a medium access control protocol in ad hoc wireless networks is the distributed arbitration for the shared channel for transmission of packets.

Routing The responsibilities of a routing protocol include exchanging the route information; finding a feasible path to a destination based on criteria such as hop length, minimum power required, and lifetime of the wireless link; gathering information about the path breaks; mending the broken paths with minimum processing power and bandwidth; and utilizing minimum bandwidth.

Multicasting Multicasting plays an important role in the typical applications of ad hoc wireless networks namely, emergency search-and-rescue operations and military communication. The arbitrary movement of nodes changes the topology dynamically in an unpredictable manner.

Transport Layer Protocol The main objective of the transport layer protocol includes setting up and maintaining end-to-end connections, reliable end-to-end delivery of data packets, flow control, and congestion control.

Self-organization One very important property that an ad hoc wireless network should exhibit is organizing and maintaining the network by itself. The major activities that an ad hoc wireless network is required to perform for self-organization are neighbor discovery, topology organization, and topology reorganization.

5.2 Related Work

There are increasing demand and importance in supporting group communications [5] over mobile ad hoc networks. Example applications include the exchange of messages among a group of soldiers in a battlefield, communications among the firemen in a disaster area, and the support of multimedia games and teleconferences with a one-to-many or many-to-many transmission pattern; multicast is an efficient method to achieve group communications. The conventional multicast protocols [6, 7] generally do not have good scalability due to the overhead of route searching, group membership management, and tree /mesh structure creation and maintenance over the dynamic topology of MANET.

These protocols are usually composed of the following three components that generally cannot scale to large network size:

Group Membership Management The management becomes harder for a large group.

Creation and Maintenance of a Tree or Mesh-based Multicast Structure These will cause significant control overhead over the dynamic topology of MANET.

Multicast Packet Forwarding The multicast packets are forwarded along the pre-built tree or mesh structure, which is vulnerable to be broken over the dynamic longer paths [8, 9].

Examples for Tree-based protocol are MAODV [10], AMRIS [11], and MZRP [12]. Examples for Mesh-based protocols are FGMP [2], Core-Assisted Mesh Protocol [13], and ODMRP [14], which are proposed to enhance the robustness with the use of redundant paths between the source and destination pair. Multicasting in MANET can be implemented in the network layer, the MAC layer, and/or the application layer. Accordingly, multicast routing protocols can be classified into three categories: Network (IP) layer multicast (IPLM), application layer multicast (ALM), and MAC layer multicast (MACLM). IPLM [1] is the most common type of multicasting used in ad hoc networks to design an efficient and reliable multicast routing protocol.

It operates on network (IP) layer that requires the cooperation of all the nodes in the network, as the intermediate (forwarder) nodes must maintain the multicast state per group. The network layer maintains the best effort unicast datagram service compared to other types that employ other layers than network layer.

Reactive Multicast Routing Protocols Traditional routing protocols such as on-demand multicast routing protocol (ODMRP) and multicast ad hoc on-demand distance vector (MAODV) [10] are reactive multicast routing protocols where the routes are created as and when required. These protocols are well suited for large-scale, narrow-band MANET with moderate or low mobility.

5.3 Protocol Description

5.3.1 Operation of Multicast Routing Protocols

Multicast routing protocols for ad hoc wireless networks are broadly classified into two types: Source-initiated Protocols and Receiver-initiated Protocols.

5.3.1.1 Source-initiated Protocols

This section deals with the events as they occur in a source-initiated protocol that uses a soft state approach. In the soft state maintenance approach, the multicast tree or mesh is periodically updated by means of control packets. In such protocols, the source of the multicast group periodically floods a JoinRequest (JoinReq) packet throughout the network. This is propagated by other nodes in the network, and it eventually reaches all the receivers of the group.

A node that wishes to join a group should respond with a *JoinReply (JoinRep)* packet, which is propagated along the reverse path of that followed by the *JoinReq* packet. In this protocol, communication between the nodes was done by propagation of join request and join reply. This is a two-phase protocol for establishing the tree (or mesh). There is no explicit procedure for route repair. In soft state protocol, the source periodically initiates the above procedure. Hard state is similar to that of a soft state source-initiated protocol, except that there is an explicit route repair procedure that is initiated when a link break is detected.

5.3.1.2 Receiver-initiated Protocols

In the Receiver-initiated multicasting protocols, the receiver uses flooding to search for paths to the sources of the multicast groups to which it belongs. In this protocol, the receiver floods a JoinReq packet which is propagated by the other nodes. Usually, the sources of the multicast group and/or nodes which are already part of the multicast tree (or mesh) are allowed to respond to the JoinReq packet with a JoinRep packet, indicating that they would be able to send data packets for that multicast group. The receiver replies with the smallest hop count and sends a *JoinAcknowledgment (JoinAck)* packet along the reverse path.

5.4 Classification of Multicast Routing Protocols

Multicast routing protocols for ad hoc wireless networks can be broadly classified into two types: Application-independent /Generic multicast protocols and Application-dependent multicast protocols. While Application-independent multicast protocols are used for conventional multicasting, Application-dependent multicast protocols are meant only for specific applications for which they are designed.

1. *Based on Topology*: In this section, based on topology multicast routing protocols are broadly classified into two types. *Tree-based* [10–12] and *Mesh-based* [2, 13, 14]. In tree-based multicast routing protocols, there exists only a single path between a source–receiver pair, whereas in mesh-based multicast routing protocols, there may be more than one path between a source–receiver pair. Tree-based multicast protocols are more efficient compared to mesh-based protocols, but mesh-based multicast protocols are robust due to the availability of multiple paths between the source and receiver. Tree-based multicast protocols can be further divided into two types source-tree-based and shared-tree-based. In *source-tree-based* multicast protocols, the tree is rooted at the source, whereas in *shared-tree-based* multicast protocols, a single tree is shared by all the sources within the multicast group and is rooted at a node referred as the core node. The source-tree-based multicast protocols perform better than the shared-tree-based protocols at heavy loads because of efficient traffic distribution. But, the latter type of protocols is more scalable. The main problem in a shared-tree-based multicast protocol is that it heavily depends on the core node, and hence, a single point of failure at the core node affects the performance of the multicast protocol.
2. *Based on initialization of the multicast session*: The multicast group formation can be initiated by the source as well as by the receivers. In a multicast protocol, if the group formation is initiated only by the source node, then it is called a source-initiated multicast routing protocol, and if it is initiated by the receiver of the multicast group, then it is called a receiver-initiated multicast routing protocol. Some multicast protocols do not distinguish between source and receiver for initialization of the multicast group.
3. *Based on the topology maintenance mechanism*: Maintenance of the multicast topology can be done either by the soft state approach or by the hard state approach. In the *soft state approach*, control packets are flooded periodically to refresh the route, which leads to a high packet delivery ratio at the cost of more control overhead, whereas in the *hard state approach*, the control packets are transmitted (to maintain the routes) only when a link breaks, resulting in lower control overhead, but at the cost of a low packet delivery ratio.

5.5 Tree-based Multicast Protocols

Tree-based multicasting is a well-established concept used in several wired multicast protocols to achieve high multicast efficiency. In tree-based multicast protocols, there is only one path between a source–receiver pair. There are two types of tree-based multicast protocols: source-tree-based multicast routing protocols and shared-tree-based multicast routing protocols. In a source-tree-based protocol, a single multicast tree is maintained per source, whereas in a shared-tree-based protocol, a single tree is shared by all the sources in the multicast group. Shared-tree-based multicast protocols are more scalable compared to source-tree-based multicast protocols. By scalability, it is meant the ability of the protocol to work well without any degradation in performance when the number of sources in a multicast session or the number of multicast sessions is increased. In source-tree-based multicast routing protocol, an increase in the number of sources gives rise to a proportional increases in the number of source-trees. Another factor that affects the scalability of source-tree-based protocols is the memory requirement. When the multicast group size is large with a large number of multicast sources, in a source-tree-based multicast protocol, the state information that is maintained per source per group consumes a large amount of memory at the nodes. But in a shared-tree-based multicast protocol, since the state information is maintained per group, the additional memory required when the number of sources increases is not very high.

5.5.1 Bandwidth-efficient Multicast Routing Protocol (BEMRP)

Ad hoc networks operate in a highly bandwidth-scarce environment, and hence bandwidth efficiency is one of the key design criteria for multicast routing protocols. Bandwidth-efficient multicast routing protocol (BEMRP) [15] tries to find the nearest forwarding node, rather than the shortest path between source and receiver. This protocol requires low communication overhead, since it does not require periodical transmission of control packets, and thus, it reduces the number of data packet transmissions. To maintain the multicast tree, it uses the hard state approach (Fig. 5.1), which is to rejoin the multicast group; a node transmits the required control packets only after the link breaks.

5.5.1.1 Tree Initialization Phase

In BEMRP, the multicast tree construction is initiated by the receiver. When a receiver wants to join a group, it initiates flooding of join control packets. The existing members of the multicast tree, on receiving these packets, respond with reply packets. When many such reply packets reach the requesting node, it chooses one of them and sends a reserve packet on the path taken by the chosen reply packet.

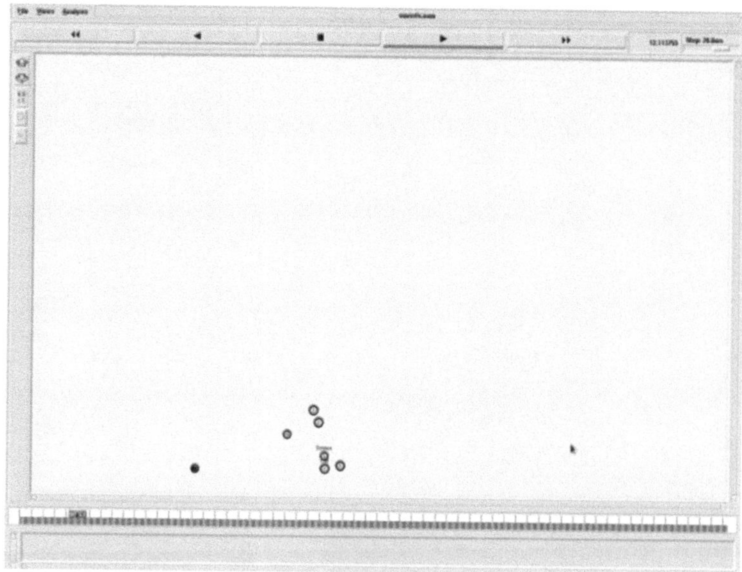

Fig. 5.1 Node formation using hard state approach

5.5.1.2 Tree Maintenance Phase

Broadcast-multicast Scheme

In this scheme, the upstream node is responsible for finding a new route to the previous downstream node (Fig. 5.2).

Local Rejoin Scheme

In this scheme, the downstream node of the broken link tries to rejoin the multicast group by means of limited flooding of the join packets.

5.5.1.3 Route Optimization Phase

When a tree node or a receiver node comes within a transmission range of other tree nodes, then unwanted tree nodes are pruned by sending the QUIT message.

The main advantage of this multicast protocol is that it saves bandwidth due to the reduction in the number of data packet transmissions and the hard state approach being adopted for tree maintenance. The main disadvantage of this protocol is, since the protocol uses the hard state approach for route repair, a considerable amount of time is spent by the node in reconnecting to the multicast session, which adds to the delay in packet delivery.

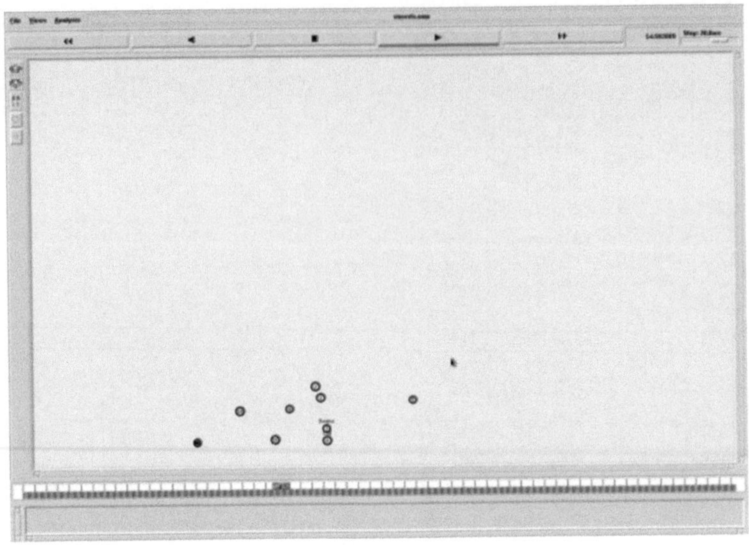

Fig. 5.2 Movement of nodes in multicast tree initialization

5.5.2 *Multicast Routing Protocol Based on Zone Routing (MZRP)*

In multicast zone routing protocol (MZRP) [12], the flooding of control packets by each node which searches for members of the multicast group (Fig. 5.3) is controlled by using the zone routing mechanism. In zone routing, each node is associated with a routing zone. For routing, a proactive approach is used inside the zone (the node maintains the topology inside the zone, using a table-driven routing protocol), whereas a reactive approach is used across zones. This protocol combines the best of both on-demand and table-driven routing approaches.

5.5.2.1 Tree Initialization Phase

To create a multicast delivery tree over the network, the source initiates a two-stage process. In the first stage, the source tries to form the tree inside the zone, and then in the second stage, it extends the tree to the entire network.

5.5.2.2 Tree Maintenance Phase

Once the multicast tree is created, the source node periodically transmits TREE-REFRESH packets down the tree to refresh the multicast tree. If any tree node does

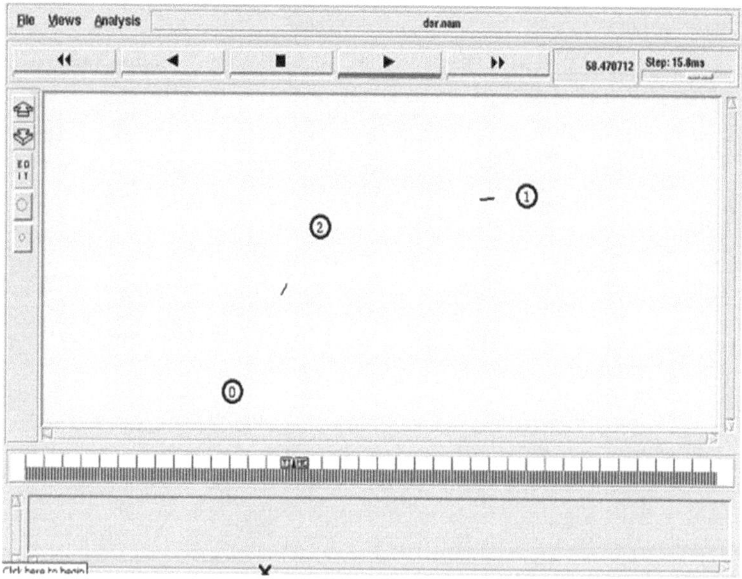

Fig. 5.3 Tree initialization in MZRP

not receive TREE-REFRESH packet within a specific time period, it removes the corresponding stale multicast route entry.

MZRP [12] has reduced control overhead as it runs over ZRP (Fig. 5.4). The fact here is unicast and multicast routing protocols can exchange information with each other. MZRP is important as it shows the efficiency of the zone-based approach to multicast routing. The size of the zone is very important in MZRP. The size should be neither too large nor too small. The main disadvantage of this protocol is that a receiver node which is located far off from the source needs to wait for a long time before it can join the multicast session.

5.5.3 Multicast Core-extraction Distributed Ad Hoc Routing (MCEDAR)

To increase the robustness while maintaining the efficiency, a different approach is used in multicast core-extraction distributed ad hoc routing (MCEDAR) [16]. A source tree over an underlying mesh infrastructure called mgraph is used for forwarding data packets. In this architecture, a minimum dominating set (MDS), which consists of certain nodes (called core nodes) in the network (Fig. 5.5), is formed using a core computation algorithm. After joining the MDS, each core node issues a piggy-backed broadcast through its beacon packet to inform its presence up to the next three hops.

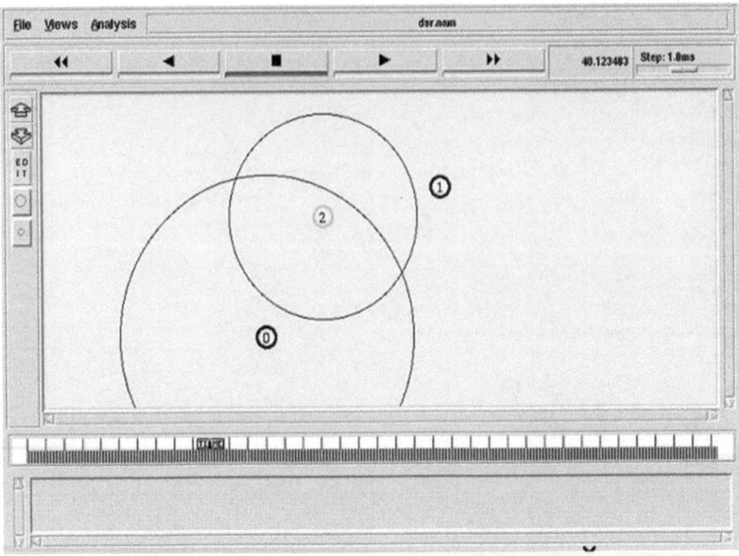

Fig. 5.4 Multicast tree maintenance in MZRP

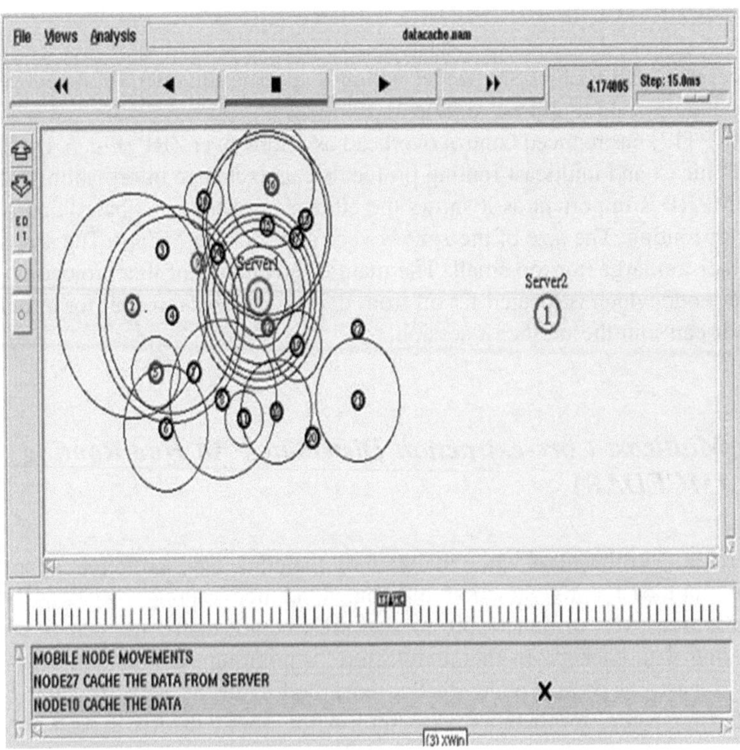

Fig. 5.5 MDS formation in MCEDAR

This multicast routing protocol is robust, and using source-tree over mesh for forwarding the data packet makes it as efficient as other tree-based multicast routing protocols. Depending on the *robustness factor* parameter, the dominator node of a receiver has multiple paths to the multicast session. So even if the current path breaks, the dominator node always has an alternate path to the multicast session. The main disadvantage of this protocol is that is more complex compared to other tree-based multicast routing protocols.

5.6 Mesh-based Multicast Routing Protocols

In ad hoc wireless networks, wireless links break due to the mobility of the nodes. In the case of multicast routing protocols, the path between a source and receiver, which consists of multiple wireless hops, suffers very much due to link breaks. Multicast routing protocols which provide multiple paths between a source–receiver pair are classified as mesh-based multicast routing protocols. The presence of multiple paths adds to the robustness of the mesh-based protocols at the cost of multicast efficiency. In this section, some of the mesh-based protocols are described in detail.

5.6.1 On-Demand Multicast Routing Protocol (ODMRP)

In the On-demand multicast routing protocol (ODMRP) [14], a mesh is formed by a set of nodes called forwarding nodes which are responsible for forwarding data packets between a source–receiver pair (Fig. 5.6). These forwarding nodes maintain the *message-cache* which is used to detect duplicate data packets and duplicate JoinReq control packets.

5.6.1.1 Mesh Initialization Phase

In this, a multicast mesh is formed between the source and the receiver. To create the mesh, each source in the multicast group floods the JoinReq control packets periodically.

Upon reception of the JoinReq control packet from the source, potential receivers can send JoinReply through the reverse shortest path. The route between a source and receiver is established after the source receives the JoinReply packets.

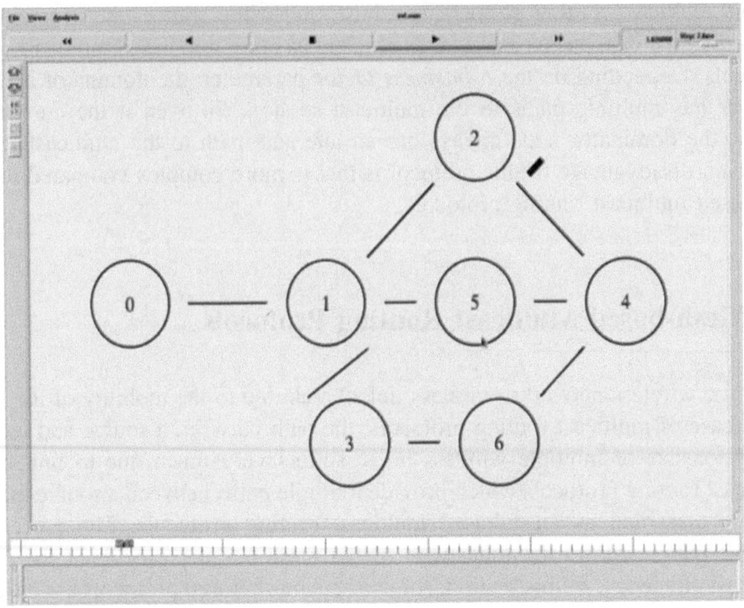

Fig. 5.6 Mesh topology in ODMRP

5.6.1.2 Mesh Maintenance Phase

In this phase, attempts are made to maintain the multicast mesh topology formed with sources, forwarding nodes, and receivers. In Fig. 5.7, if path between 1 and 2 fails, then it finds the alternate path to reach the destination 4 (Fig. 5.8).

Since ODMRP uses the soft state approach for maintaining the mesh, it exhibits robustness. But, this robustness is at the expense of high control overhead. Another disadvantage is that the same data packet (from source to receiver) propagates through more than one path to a destination node (Figs. 5.7 and 5.8), resulting in an increased number of data packet transmissions, thereby reducing the multicast efficiency.

5.6.2 Dynamic Core-based Multicast Routing Protocol (DCMP)

The dynamic core-based multicast routing protocol (DCMP) [3] attempts to improve the efficiency of the ODMRP multicast protocol by reducing control overhead and providing better packet delivery ratio. Mesh-based protocols, such as ODMRP, suffer from the disadvantages:

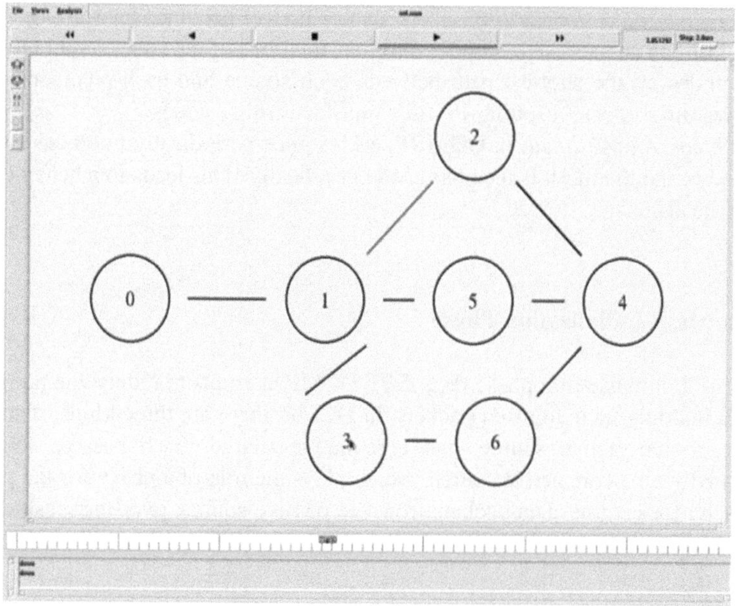

Fig. 5.7 Route failure in ODMRP

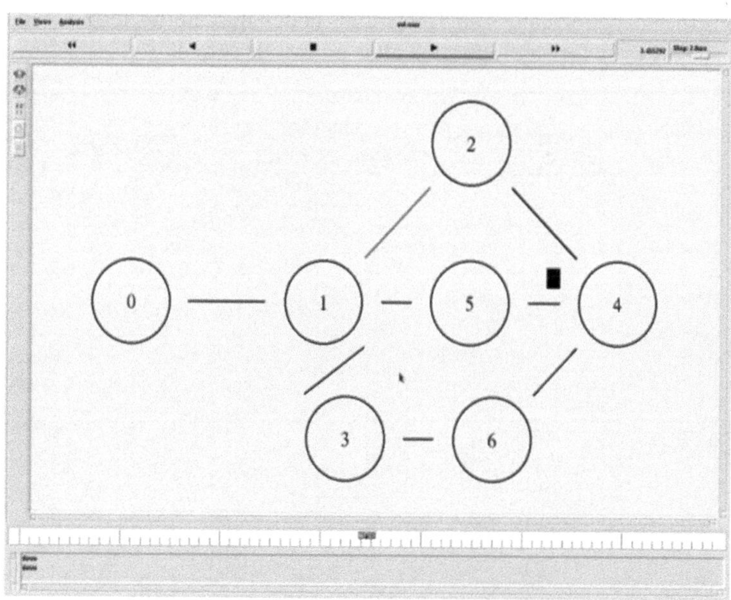

Fig. 5.8 Finding alternative path in ODMRP

1. *Excessive data forwarding*: Too many nodes become forwarding nodes, resulting in an excessive number of retransmissions of data packets. In ODMRP, all nodes on the shortest path between each source and each receiver become forwarding nodes, resulting in too many forwarding nodes.
2. *High control overhead*: In ODMRP, each source periodically floods its JoinReq packets and the mesh is reconstructed periodically. This leads to a heavy control overhead.

5.6.2.1 Mesh Initialization Phase

In the mesh initialization phase (Fig. 5.9), DCMP attempts to reduce the number of sources flooding their JoinReq packets. In DCMP, there are three kinds of sources: passive source, active source, and core active source. Each passive source is associated with a core active source, which plays the role of a proxy for the passive source, by forwarding data packets from the passive source, over the mesh created by its JoinReq packets.

5.6.2.2 Mesh Maintenance Phase

DCMP's mesh maintenance is soft state, similar to that of ODMRP. Thus, the mesh is reconstructed periodically and forwarding nodes that are no longer part of the mesh, cancels their forwarding status after a timeout period.

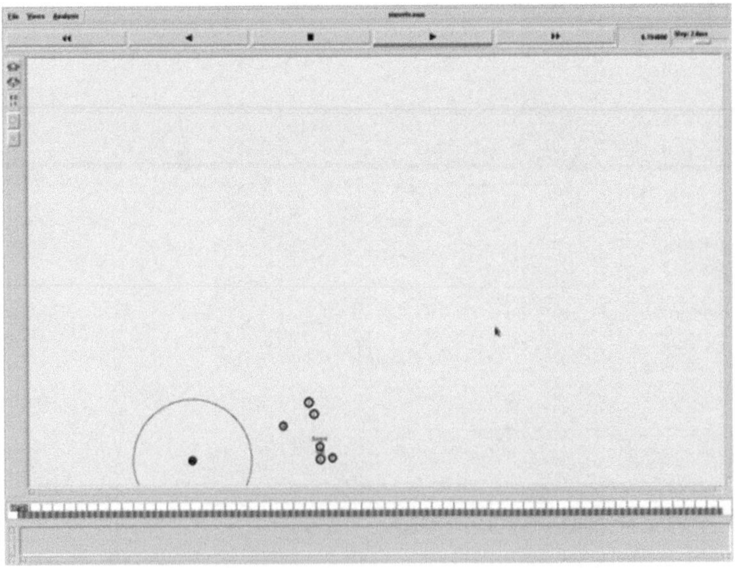

Fig. 5.9 Mesh initialization phase in DCMRP

The primary advantage of DCMP [3] is its scalability due to decreased control overhead and its superior packet delivery ratio. The performance improvement of DCMP over ODMRP increases with the number of sources in the multicast. One of the drawbacks of DCMP is that the parameters associated with it, *MaxPassSize* and *MaxHop*, are likely to depend on the network load conditions, group size, and number of sources, and optimal values of these parameters may even vary from one node to another.

5.6.3 Forwarding Group Multicast Protocol (FGMP)

Another mesh-based multicast routing protocol [2], which is also based on the forwarding group concept, is forwarding group multicast protocol (FGMP-RA [Receiver Advertising]). The major difference between ODMRP and FGMP-RA lies in who initiates the multicast group formation. ODMRP is a source-initiated multicast routing protocol, whereas FGMP (FGMP-RA) is a receiver-initiated multicast routing protocol.

5.6.3.1 Mesh Initialization Phase

In order to form the multicast mesh, each receiver floods the JoinReq control packet in the network (Fig. 5.10). When the source receives these JoinReq control packets, each source updates its *member table* and creates a *forwarding table*. The member

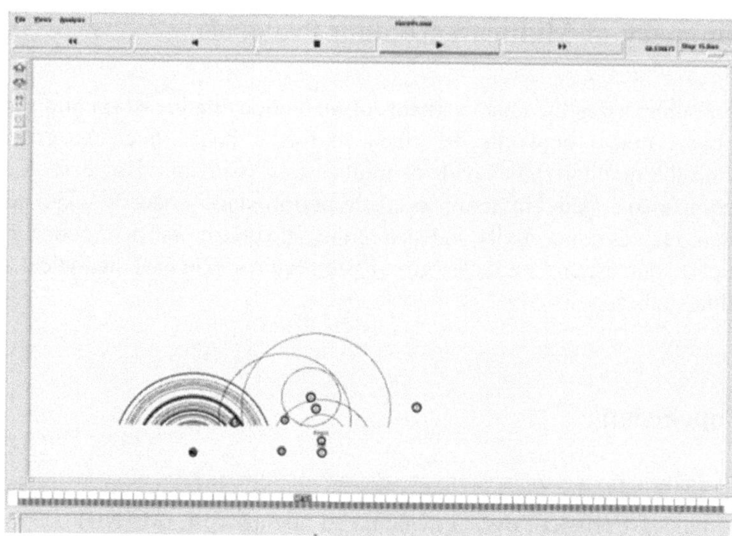

Fig. 5.10 Group formation in FGMP

Table 5.1 Summary of multicast routing protocols

Multicast protocols	Multicast topology	Initialization	Maintenance approach
BEMRP	Source-tree	Receiver	Hard state
MCEDAR	Source-tree over mesh	Source or receiver	Hard state
MZRP	Source-tree	Source	Hard state
ODMRP	Mesh	Source	Soft state
DCMP	Mesh	Source	Soft state
FGMP	Mesh	Receiver	Soft state

table keeps the ID's of all the receivers in the multicast group. After creating the forwarding table, the source forwards this table toward the receivers. When the forwarding table reaches the receivers, the route between the source–receiver pairs get established.

5.6.3.2 Mesh Maintenance Phase

In this, the movement of the receiver, the route breaks but receiver can still receive data packets through another. To maintain a route, FGMP uses the soft state approach, that is, receivers periodically flood the JoinReq packet to refresh the route.

Due to its mesh topology and soft state maintenance scheme, it is more robust as compared to the tree-based multicast routing protocols.

5.7 Summary of Multicast Routing Protocols

Table 5.1 summarizes the characteristics of the various multicast routing protocols for ad hoc wireless networks described so for. It helps in characterizing and identifying the qualitative behavior of multicasting protocols. However, a quantitative comparison study in terms of their performance under a wide range of parameters such as connectivity and size of the network, mobility of nodes, number of multicast sources, and multicast group size requires extensive analytical and /or simulation studies.

5.8 Conclusion

There is an increasing demand and a big challenge to design more scalable and reliable multicast protocol over a dynamic ad hoc network (MANET). This paper presents a comprehensive survey on efficient multicast routing protocols namely

BEMRP, MZRP, MCEDAR, ODMRP, DCMP, and FGMP which are used to achieve Group Communication. Working nature of these multicast routing protocols have been analyzed for improving the performance and to achieve group communication in an efficient manner. Substantial research efforts over the last decade have been focused on developing and implementing routing protocols and security techniques that better suited the nature of MANETs.

References

1. Mo'men, A.M.A., Hamza, H.S., Saroit, I.A.: A Survey on Security Enhanced Multicast Routing Protocols in Mobile Ad Hoc Networks, ISBN 978-1-14244-9924-3. IEEE (2010)
2. Chiang, C.C., Gerla, M., an Zhang, L.: Forwarding group multicasting protocol for Multi-Hop, mobile wireless networks. Cluster Comput **1**(2), 187–196 (1998). Special Issue on Mobile Computing
3. Das, S.K., Manoj, B.S., Siva Ram Murthy, C.: A dynamic core-based multicast routing protocol for Ad Hoc mobile networks. Proceedings of the 3rd ACM International Symposium on Mobile Ad Hoc Networking & Computing (MobiHoc '02), pp. 24–35, 2002
4. Chao G., Mohapatra P,: Scalable Multicasting in Mobile Ad Hoc Networks. ISBN 0-7803-8355-9, IEEE (2004)
5. Xiang, X., Wang, X., Yang, Y.: Supporting efficient and scalable multicasting over mobile ad hoc networks. IEEE Trans. Mobile Comput. **10**(5), 544–559 (2011)
6. Royer, E.M., Perkins, C.E.: Multicast operation of the ad hoc on demand distance vector routing protocol. In CM/IEEE MOBICOM, pp. 207–218, 1999
7. Gerla, M., Lee, S.J., Su, W.: On-demand multicast routing protocol (ODMRP) for ad hoc networks. Internet draft, draft-ietfmanet-odmrp-02.txt (2000)
8. Kar, B., Kung, H.T.: Greedy perimeter stateless routing for wireless networks. ACM/IEEE MOBICOM, pp. 243–254, (2000)
9. Kuhn, F., Wattenhofer, R., Zhang, Y., Zollinger, A.: Geometric ad-hoc routing: of theory and practice. In: International Symposium on the Principles of Distributed Computing (PODC), 2003
10. Li, J., et al.: A scalable location service for geographic ad hoc routing", in ACM/IEEE MOBICOM, pp. 120–130, 2000
11. Wu, C.W., Tay, Y.C., Toh, C.K.: Ad hoc multicast routing protocol utilizing increasing id-numbers (AMRIS) functional specification. Internet draft, draft-ietf-manet-amris-spec-00.txt, 1998
12. Devarapalli, V., Selcuk, A.A., Sidhu, D.: MZRP: a multicast protocol for mobile ad hoc networks. Internet draft, draft-vijayf-manet-mzr-01.txt (2001)
13. Garcia-Luna-Aceves, J.J.: Madruga, E.L: the core-assisted mesh protocol. IEEE J-SAC, **17**(8), 1380–1994 (1999)
14. Lee, S.J., Gerla, M., Chiang, C.C.: On-demand multicast routing protocol. In: Proceedings of IEEE WCNC 1999, pp. 1298–1302, 1999
15. Ozaki, T., Kim, J.B., Suda, T.: Bandwidth efficient multicast routing protocol for ad hoc networks. Proceedings of IEEE ICCCN 1999, pp. 10–17, 1999
16. Sinha, P., Sivakumar, R., Bharghavan, V.: MCEDAR: multicast core extraction distributed ad hoc routing. Proceedings of IEEE WCNC 1999,pp. 1313–1317, 1999

Chapter 6
Token-based Group Local Mutual Exclusion Algorithm in MANETs

Ashish Khanna, Awadhesh Kumar Singh and Abhishek Swaroop

Abstract In this paper, a generalization of the group mutual exclusion problem based upon the concept of neighborhood has been proposed and named as group local mutual exclusion (GLME). The neighborhood is defined based upon the location of shared resources and no synchronization is required between nodes in two different neighborhoods. A token-based solution of the GLME has also been proposed. The algorithm satisfies safety, starvation freedom, and concurrent occupancy properties. The proof of correctness has also been included in the present paper. To the best of our knowledge, it is the first token-based algorithm to solve GLME problem in MANETs.

Keywords Ad hoc network · Local mutual exclusion · Neighborhood · Resource allocation

6.1 Introduction

The efficient and effective use of resources is a highly desirable property in MANETs as the environment is inherently resource constrained. Due to dynamic topology, limited bandwidth, limited battery power, and low processing capability, algorithms designed for synchronization in static networks cannot be directly applied in MANETs. Therefore, several algorithms for various resource allocation

A. Khanna (✉)
MAIT, Delhi 110085, India
e-mail: ashishk746@yahoo.com

A.K. Singh
NIT, Kurukshetra 136119, Haryana, India
e-mail: aksinreck@rediffmail.com

A. Swaroop
Galgotias University, Greater Noida, U.P., India
e-mail: asa1971@gmail.com

S. Patnaik et al. (eds.), *Recent Development in Wireless Sensor and Ad-hoc Networks*, Signals and Communication Technology, DOI 10.1007/978-81-322-2129-6_6

problems, e.g., mutual exclusion [1], *k*-mutual exclusion [2], group mutual exclusion (GME) [3] have been proposed in the literature. There are two major approaches to handle mutual exclusion problem and its variants, namely token-based and nontoken-based, a.k.a. permission-based [3].

The classic mutual exclusion problem [1] is to ensure that only one of the contending nodes is allowed to enter critical section (CS) at a time. The group mutual exclusion problem proposed by Joung [4] is an interesting variant of classical mutual exclusion problem. The GME problem [4] deals with two fundamentally opposite issues, namely mutual exclusion and concurrency. In GME, each request is associated with a type or group. The processes requesting the same group are allowed to be in their CS simultaneously. However, the processes requesting different groups must execute their CS in mutually exclusive way. GME problem has been modeled as congenial talking philosopher (CTP) problem by Joung [5]. In CTP problem, there are n philosophers and m forums; however, there is only one meeting room. A philosopher may be in any one of the following three states: (i) thinking (ii) waiting and (iii) talking in a forum. A philosopher interested in a forum may enter the meeting room, if the meeting room is empty or some philosophers interested in the same forum are already in the meeting room, otherwise he has to wait.

An example of GME problem is a situation in which large amount of data stored in some secondary storage device (such as CD jukebox) is being shared by several processes. The processes are interested in accessing some data stored on the CD's. However, due to the limited amount of buffer space available, only one CD can be loaded in the buffer at a time. Therefore, only the processes, interested in the data stored in the currently loaded CD, may access the required data concurrently.

Attiya et al. [6] introduced the concept of local mutual exclusion for MANETs. In local mutual exclusion, no two neighboring nodes can enter CS simultaneously; however, the nodes which are not neighbors can be in their CS simultaneously. On the other hand, in classical or global mutual exclusion, no two nodes (how far apart) can be in CS simultaneously. Kogan [7] claims that in MANETs, the local mutual exclusion problem has more potential applications in comparison with global mutual exclusion problem. Khanna et al. [8] proposed *k*-local mutual exclusion problem (KLME) in MANET's and solved it using a token-based approach. Luo et al. [9] defined a new problem named local group mutual exclusion (LGME) as a variant of group mutual exclusion problem specially suited for VANET's and proposed a coterie-based solution for LGME. However, their concept of conflicting nodes is different from the neighborhood concept used in KLME.

In KLME, the neighborhood has been defined as the independent smaller unit of a larger geographical area. The larger area is divided in smaller neighborhoods based upon the location of shared resources. Each set of shared resources placed at one location can be accessed in a particular neighborhood only. In this paper, the group local mutual exclusion (GLME) problem with the concept of neighborhood (similar to [8]) has been handled using a token-based approach. An interesting application of the problem may be the situation in which multiple replicas of a read only database are available at different locations. At each location, due to limited amount of buffer, only a subset of database may be in the buffer and the processes (in the range of the current replica) requesting the same portion of database may be in CS simultaneously, however, the processes requesting the different portion have to wait. Moreover, the processes in the range of another replica may be in CS using a different portion of the database.

Any solution to the GLME problem must satisfy following properties:

Mutual exclusion: No two processes in the same neighborhood requesting different groups (resources) can be in their CS simultaneously.

Bounded delay: A process attempting to enter CS will eventually succeed.

Concurrent entering: If some processes in the neighborhood are interested in entering CS using a group and no process in the neighborhood is interested in a different group then the processes can concurrently enter CS.

The rest of the paper has been organized in the following manner. Section 6.2 contains the related work, and Sect. 6.3 contains the system model and working of the proposed algorithm. Section 6.4 contains the performance analysis of the algorithm and in Sect. 6.5 proof of correctness has been discussed. Finally, Sect. 6.6 concludes the work and contains future research directions.

6.2 Related Works

The *dining philosopher's problem* [2] is an interesting variant of mutual exclusion problem in static networks. Although, the solution to dining philosopher's problem works as a solution to mutual exclusion problem, it has a special property that the failure of a node does not affect the entire system. The resource allocation problem is also an active area of research in MANETs. Recently, Sharma et al. [10] presented a detailed survey of the mutual exclusion in MANETs. The dining philosopher's problem has been extended by Attiya et al. [6]. Kogan [7] explained that unlike conventional mutual exclusion, local mutual exclusion algorithms have better failure locality. In local mutual exclusion, no two neighboring nodes can enter CS simultaneously; however, the nodes which are not neighbors can be in their CS simultaneously. On the other hand, in classical or global mutual exclusion, no two nodes (how far apart) can be in CS simultaneously. Khanna et al. [8] presented a KLME using token-based approach in MANET's. In KLME, the concept of neighborhood is based upon the location of sets of resources scattered in the entire area, all the nodes in the range of one set of resources comprises one neighborhood. The nodes are mobile and hence can move from one neighborhood to another. Wu et al. [9] defined a new problem named LGME as a variant of group mutual exclusion problem specially suited for VANET's and proposed a coterie-based solution for LGME. However, they took the example of automatic vehicles and considered the traffic lanes as resources. The vehicles with crossing paths are considered conflicting. On the other hand, the vehicles with non-crossing paths are nonconflicting and no synchronization is required among these. The concept of local versus nonlocal used in LGME is different from the concept of neighborhood based upon resource location used in KLME. In the present paper, a variant of group mutual exclusion problem using the concept of neighborhood [8] has been proposed and solved using a token-based approach. Although, the problem defined in this paper appears to be similar to the problem defined by Wu et al. [9], there are some following major differences. (i) In GLME, a node may move to any neighborhood from its current neighborhood, whereas in LGME [9], a node (vehicle) may not change lane arbitrarily. Therefore, the environment in GLME is more dynamic in comparison with the environment in LGME [9]. (ii) The neighborhoods in our problem are nonoverlapping, and in each neighborhood, one group may be active, whereas in LGME, there are conflicting lanes and nonconflicting lanes and nonconflicting lanes cannot be activated simultaneously. (iii) In LGME, all resources are lanes, whereas in GLME, the number of resources is multiple and the resources may be of different types. Hence, GLME can be used in various types of applications. Recently, a lot of research is going on mutual exclusion problem and its variants [11–14] in MANETs; however, the problem of local mutual exclusion and its variants are still unexplored.

6.3 System Model

We consider a MANET, in which, there are n mobile nodes and the resources to be shared are located at different places. The area in which a particular set of resources can be accessed defines a neighborhood. The mobile nodes in a neighborhood compete for the resources available in the neighborhood.

1. There is no assumption over the processing speed of a mobile anode in the ad hoc environment.
2. The delay in message propagation is finite but unpredictable.
3. A mobile node cannot make a new request until its old request is satisfied.
4. A workable link layer protocol is in place.
5. No node will leave its neighborhood area while in CS. If a node executing in CS wants to exit the neighborhood, it will come out of CS before leaving the neighborhood.
6. Underlying ad hoc network is of quasi stable nature.
7. Message delivery is ensured between two connected nodes via a routing protocol.
8. Each node has unique identifier.
9. Number of nodes in the system is finite.

The similar assumptions have been made by previous works on MANETs [8].

6.3.1 Working

In this section, brief working of our algorithm is discussed in MANETs. To the best of our knowledge, this is the first token-based solution for GLME problem in MANET's.

In this algorithm, n numbers of nodes are considered in local neighborhood. After initialization, node 1 is elected as token holder; and token information is broadcasted, so that neighboring nodes may transfer their request to token holder.

When a node i receives a non-stale token information after the updating process, if the node i is in waiting state with the same group of request, and if node i is part of tokens follower list, it enters CS. Otherwise, it will send the request for CS to the sender. In case, node i is in remainder section and is part of token's follower list, it will send leaving CS message.

When any node requests for CS, there are following possibilities. (a) Node is idle token holder node. It enters CS. (b) Token holder but not in CS and its group type matches the current group. The node enters CS. (c) Token holder, not in CS, and its group type does not match the current group. The request is added in token queue. (d) The node does not have token information. It will wait for token information. (e) The node has token information. It will forward request to token holder.

When the token holder receives a non-stale request, request acknowledgment is sent to the token holder and depending upon its state further actions is performed on the receiving node. In case a nontoken holder node receives request, request is stored in the local list of node.

When node i receives token after the updating process, if it is in waiting state with matching group of request, it will enter CS and will send permission to enter CS to all the requesting processes in token queue with matching group. The node i will take further actions shown in pseudo code. A node receiving permission to enter CS as follower will enter CS.

A node receiving non-stale request acknowledgment will set the request acknowledgment flag. A node receiving permission to enter CS as follower will enter CS.

In case, a token holder node i exits from CS, if number of followers are zero and token queue is empty, its state is changed to holding idle token. Otherwise, if token queue is not empty, the token is transferred to the node at front of token queue and other nodes having requests for matching group in token queue are also permitted to enter CS. However, if followers are still in CS, token holder state is changed accordingly. In case a nontoken holder node comes out of CS, it sends release CS message to token holder.

When token holder receives a release message from any follower node, follower node is removed from token follower's list. In case, token holder is not in CS and it is holding token with some followers, appropriate actions are taken as shown in pseudo code. In case, a nontoken holder receives release CS from any follower, this information is stored in the local data structure of receiving node for future action.

When a node is about to leave the neighborhood, various actions are performed depending upon its state. In case leaving node is in CS, its state is changed to waiting state which not happens in generalized GME algorithm and is a striking feature of GLME algorithm. When any node receives the leaving information of outgoing node, receiving nodes data structure is updated accordingly. Any new node when joins neighborhood, it is initialized and its arrival information is broadcasted. When any node receives this arrival information, receiving node updates the new node into its data structure and a token holder node receiving this information sends token information to the new node.

6.3.2 Data Structure

Data structure at node i ($1 \leq i, j \leq n$)

state$_i$: The current state of node *i*: REM: remainder, W: waiting, HI: holding idle token, HF: holding token with followers but not in CS, CF: CS as follower and CT: CS as a token holder.

n_foll_leave$_i$: Stores information of foll_leaves_CS message when node is nontoken holder.

token_id$_i$: Stores the node id of token holder or Ø.

RN$_i$: Request number of node *i* which is incremented when node *i* requests for CS.

g_node_type$_i$: Type of group for which node has requested.

node_req_list$_i$: Stores the request number and its corresponding group number of neighboring nodes.

node_leader_no$_i$: Stores the number of times the leader/token holder has changed.

req_ack$_i$: Boolean variable indicating request acknowledgment is received.

n: Total number of nodes in neighborhood.

Data Structure at token holder

T_Q: A FIFO queue which stores nodes requests with corresponding group and request number.

gp_t_type: Stores the group being accessed by token holder in current session.

no_t_follower: Number of nodes currently in CS as follower.

t_foll_list: Node ids to which per_grant message has been sent and release from CS message from that nodes has yet not been received.

LN: An array to store the sequence number of latest served request of every node.

token_leader_no: Stores the number of times the leader/token holder has changed.

6.3.3 Messages

1. *token_info*$_i$ (*j*, *t_foll_list*, *token_leader_no*, *gp_t_type*): Node *i* broadcasts token information along with other useful information.
2. *req_CS*$_i$ (*i*, g_i,RN_i, *Token_id*$_i$): Node *i* send the request message to the token holder.
3. *token* (*j*, *T_Q*, *gp_t_type*, *no_t_followers*, *t_foll_list*, *LN*): Node receiving token becomes token holder.
4. *req_ack* (*j*, LN_i): Token holder node sends this message to the node from which it has received request for CS as an acknowledgment of request.
5. *per_grant* (g_x, *j*): Token holder node sends this to node *j* to allow it to enter CS as follower.
6. *foll_release_CS* (*j*, RN_i): Sent by follower node to token holder node after coming out of CS.
7. *I_am_leaving* (*i*, *n_foll_leave*$_i$, *node_req_list*$_i$): Broadcasted by leaving node.
8. *I_have_join* (*i*): Broadcasted by node entering neighborhood.

6.3.4 Pseudo Code of Algorithm at Node i

```
a) Initialization
g_node_typei=Ø; node_leader_noi=0; req_acki=false
for(i=1 to n) statei=REM; token_idi=Ø; RNi=0;
n_foll_leavei=Ø; g_node_typei=Ø;
for{i=1 to n}
   for{j=1 to n} node_req_listi[j]={0, Ø}
TokenGenerate()
 Let node 1 is initially elected as token Holder
 token_id=1; T_Q=Ø; token_leader_no=token_leader_no + 1
 node_leader_noi=token_leader_no; gp_t_type=Ø
 no_t_followers=0; t_foll_list=Ø
 for(i=1 to n) LN[i] = 0
   token_id1=1; Broadcast token_info

b) Node i receives token information from node j
if (token_leader_no>node_leader_noi)
 token_idi=j; node_leader_noi=token_leader_no
  if (statei=W && g_node_typei=gp_t_type)
    if(i ε t_foll_list)enter CS;call exit_CS;rec_acki=true
    else  send req_CS        // rec_acki=false
```

```
 if (state_i=REM && i ε t_foll_list)send leave_CS to j
else reject the message

c) Node_i request for CS with group g_i
 state_i=W; req_ack_i=0 ; RN_i=RN_i+1; g_node_type_i=g_i
 add request in node_req_list_i
 if(token_id_i=i)
   if(state_i=HI) enter CS; call exit_CS; state_i=CT
   if(state_i=HF && gp_t_type=g_i)
     state_i=CT; enter CS; call exit_CS
 if(token_id_i=Ø) wait for token info
 else send req_CS to token_id_i

d) node i receives req_CS from node j
if (node_req_list_i[i]<RN_j )
 if(token_id_i=i)
   send req_ack to node j; node_req_list[i]=RN_j
    if(j ε t_foll_list) remove i from t_foll_list
    if(state_i=CT)
       if(gp_t_type=g_i) LN[j]=RN_j; no_t_followers ++
         send per_grant to j
    if(state_i=HF)
       if(gp_t_type=g_i &&  T_Q=Ø)
         LN[j]=RN_j; send per_grant to j;
         no_t_followers++
    if(state_i=HI) token_leader_no+=1;
      node_leader_no_i=token_leader_no; token_id_i=j
      Send token to node j; Broadcast token_info
    else append the request in T_Q
    else store the request in node_req_list_i
   else reject the message

e) node i receives token
  Remove nodes in n_foll_list_i from t_foll_list; Decrement
  number of follower correspondingly; update T_Q with
  node_req_list_i; token_id_i=i
  if(i ε T_Q) remove i from T_Q
  if (state_i=W)
    if(gp_t_type=g_i)
      state_i=CT; enter CS; call exit_CS;
      for(k=1 to n )
        if(k ε T_Q && g_k=gp_t_type)
          send per_grant to node k; no_t_followers++
          add k in t_foll_list
        else
          if(no_t_foll=0) state_i=CT; gp_t_type=g_i
            for(k=1 to n )
              if(k ε T_Q && g_k=gp_t_type)
```

```
                Send per_grant to node k; no_t_followers ++
                add k in t_foll_list
      if(statei=CF) statei=CT; no_t_follower --;
         remove i from t_foll_list
      if(statei=REM)
          if(T_Q=Ø) statei=HI
          else send token to node(X) at front at T_Q
           token_idi=X; gp_t_type=gx; Broadcast token_info
           for(k=1 to n)
             if(k ε T_Q && gk=gp_t_type)
              Send per_grant to node k; no_t_follower ++

f) node i recevives req_ack(j,LNi)
   if(RN[i]=LNi) req_acki=true

g) nodei receives per_grant(gx,j) from node j
   statei=CF; enter CS; call exit_CS

h) nodei exit from CS
   if(token_idi=i)
    if(t_foll_list=0)
      if(T_Q = Ø) statei=HI
      else token_leader_no +=1
      node_leader_noi=token_leader_no; statei=REM
        if(X ε front of T_Q)
          gp_t_type= gx; remove X from T_Q
          for(k=1 to n )
            if(k ε T_Q && gk=gx)
               Send per_grant to k; no_t_followers ++;
               add K to t_foll_list; Send token to x;
               Broadcast token_info()
     else statei=HF
  else statei=REM; send foll_release_CS to token_idi

i) node i receives release of CS message from follower.
if(token_idi=i)
  if(LNi=RNi) remove j from t_foll_list;no_t_followers--
    if(statei=CT) do nothing
   if(statei=HF)
      if (no_t_followers=0)
        if(T_Q=Ø) statei=HI
       else let x is node at front of T_Q; gp_t_type=gx
         remove X from T_Q
          for(k=1 to n)
           if(k ε T_Q && gk=gx)
            send per_grant to k;no_t_followers++;
            add k in t_foll_list;Send token to X;
            broadcast token_info
```

```
          else do nothing     //no_t_follower ≠ 0
      else reject the message
   else Store foll_release_CS info n_foll_leave list

j) node i is about to leave the neighborhood
   if(token_id_i=i)
     if(state_i=CT or state_i=HF)
       if(state_i=CT)     state_i=W
       else     state_i=REM
        if(no_t_foll≠0)
          token_leader_no += 1;
          node_leader_no_i=token_leader_no
          token_id_i=X (lowest id node of t_foll_list)
          broadcast token_info; Send token to X
         else
          if(T_Q≠Ø)
            Send token to node at front of T_Q; store
             similar nodes having similar groups of
             request into t_foll_list & send per_grant to
             respective nodes; broadcast token_info
          else send token to lowest id in neighborhood
    if(state_i=HI)
        send token to lowest id in neighborhood
    else
        if(state_i=CF)
          send foll_leave_CS to token_id; state_i=W
        if(state_i=W)  state_i=W
   else do nothing
   broadcast I_am_leaving message

k) node i receives I_am_leaving from j
   remove leaving node j from all lists
   update n_foll_leave_j , node_req_list_j received
   if(token_id_i=i) Delete j from token's data structure

l) new node i joins neighborhood
    send I_have_join to all node in neighborhood.
    wait for token_info; apply initialization to node i

m) node i receives I_have_join
   add entry of node j into local data structure
   if(token_id_i=i)  send token_info to node j
```

6.4 Performance Analysis

6.4.1 Message Complexity

According to proposed algorithm, when the token holder node wants to enter CS, the requesting token holder node can directly enter CS without any message(s) exchanged in the best case.

When a new session has to be started, the node at the front of the token queue is selected as new leader and its group is the group for the next session.

Let there are k nodes which enter CS as follower during this session.

The number of messages exchanged for these $k + 1$ CS executions will be:

n *(broad cast new leader)* + $(k + 1)$ *(request)* + $(k + 1)$ *(request acknowledgment)*+ k *(permission grant)* + k *(follower_leave_CS)*.

Hence, *number of messages/CS* = $\{(n) + (k + 1) + (k + 1) + k + k\}/(k + 1)\}$.

The worst case will occur when there is no follower that is $k = 0$, and in this case, the messages/CS will be $n + 2$, i.e., $O(n)$.

6.4.2 Synchronization Delay

The synchronization delay is the time interval between the end of one session and start of next session. The synchronization delay is considered under heavy load conditions. Under heavy load, when one session finishes, there are always some pending requests in the token queue and the node at the front of the token queue will be selected as the new token holder. Now, there are two possibilities. (i) The new token holder is the same node which was the token holder in previous session. In this case, there will be no synchronization delay and the node will start the new session immediately. (ii) The new token holder is different from the previous token holder. The token_info message will be broadcasted and the new token leader will enter CS on receiving this message. Hence, the synchronization delay in this case will be T (where T is the maximum message propagation delay).

6.4.3 Waiting Time

The waiting time is considered under light load condition. In this case, if a node holding idle token wishes to enter CS, it will immediately enter CS. However, if the requesting node is not having token, it will forward its request to the token holder (holding idle token as the light load is being considered). On receiving this message, the node holding idle token will transfer the token to the requesting node. On receiving the token, the requesting node will enter CS after $2T$ time (T for request and T for Token transfer).

6.5 Proof of Correctness

6.5.1 Safety

Lemma 1 *There exits at most one token in the system at any point of time.*

Proof The proof trivially follows. According to our assumption, initially, there is only one token in the system. Moreover, when a token holder node leaves the neighborhood, it passes the token to its selected successor. Also we have assumed that the channel is reliable, hence, the new successor will receive the token eventually. Thus, there would always exits single token in the system. □

Theorem 1 *If more than one node is in CS simultaneously, they use the same group.*

Proof Assume the contrary. Say, two nodes N_i and N_j are in CS simultaneously and use group g_i and g_j, respectively, where $g_i \neq g_j$. Thus, the state of both nodes can be either CT or CF. Now, there are four cases possible: □

Case 1 $state_i = state_j =$ CT. From *Lemma 1,* it is clear that both nodes cannot hold token simultaneously. Thus, this case represents an infeasible system state.

Case 2 $state_i = state_j =$ CF:

This implies that none of them holds token and both the nodes have entered CS as followers. Now, there are two possible cases. (a) The node N_i and N_j received permission from the same node (say N_k). Now, in a session, a token holder node can permit the nodes having the same group and at a time only one session may be active, therefore, N_i and N_j must be using the same group. (b) The node N_i received permission from node N_k (say before N_j) and node N_j received permission from node N_m. This case is only possible when node N_k has left the neighborhood after permitting N_i (with group g_1) and sent the token to some other node. Now, the token may be forwarded to multiple nodes before it reaches N_m, however, the group in use must still be g_1 since N_i is still in CS using g_1 as follower and new session cannot be initiated till all the followers have come out of CS. Hence, node N_m must have sent the permission to N_j to enter CS with group g_1 only.

In both the cases, it has been proved that N_i and N_j are in CS using the same group which is a contradiction.

Case 3 $state_i =$ CT and $state_j =$ CF

Clearly, in this case, node N_i is the token holder. From *Lemma 1*, node N_j cannot have token simultaneously. However, node N_j is in CS (say using group g_1) and hence node N_j must have entered CS as the follower. There are following two possibilities. (a) N_j received permission to enter CS from N_i. The token holder node is in CS and it may permit only those nodes requesting the same group. Thus, node

N_i and N_j do not use different groups. (b) N_j received permission from some other node say N_k and the token is transferred to N_i later on. Now, since N_j is still in CS as follower, the session (group g_1) in which N_j entered in CS cannot be terminated and N_i must also be in CS using group g_1.

Case 4 $state_i$ = CF, $state_j$ = CT

Same as *Case 3*.
Therefore, the theorem holds.

6.5.2 Starvation Freedom

Lemma 2 *Any request message sent by a node will eventually reach the token holder node.*

Assume the contrary that a request generated by node N_i never reaches the token holder node N_j. Now, the request may not reach the token holder node if the token holder node moves out before receiving the request. However, the token holder node broadcasts information about the new token holder before transferring the token and node N_i will continue to send the request to the new token holder node till the time node N_i receives request acknowledgment from the token holder node. In this scenario, the assumption is only possible when the token is transferred each time before receiving the request of node N_i. However, the request will be stored in the local request queue of the request receiving node. Additionally, whenever a node receives token, it updates token queue with its local request queue. In worst case, this situation may be continuously repeated. However, according to our assumption, there are finite number of nodes in the system, hence, in maximum $O(N)$ (where as N is the number of nodes in the system) rounds, the token will reach a node which has already stored request of node N_i in its local request queue and the request of node N_i will be added in the token queue. This contradicts our assumption and *Lemma 2* holds.

Theorem 2 *Every request will eventually be served.*

From *Lemma 2*, it is ensured that each request will reach the token holder node. It has been assumed that each process will remain in CS for finite time. Moreover, in case of a conflicting request available in token queue, the token holder stops sending permission to processes requesting the group currently being used. Hence, any session will eventually terminate in the presence of conflicting request (s). Additionally, the token queue is *FCFS*; hence, any request will eventually be served. Thus, *Theorem 2* is proved.

6.5.3 Concurrent Occupancy

Theorem 3 *In the absence of conflicting request(s), any process requesting for currently active group will be permitted to enter CS.*

From *Lemma 1*, it is evident that the request of node N_i (requesting group g_1, which is currently active) will eventually reach the token holder node. Moreover, the token holder node checks the token queue. Since, there are no conflicting requests; the token queue must be empty and the token holder will send permission to node N_i and node N_i will enter CS as follower.

6.6 Conclusions and Future Scope

This paper is inspired by the fact that GLME has multiple applications in MANETs. The present paper proposes a token-based GLME algorithm for MANETs. This algorithm can also be used in multiple neighborhoods instead of using a conventional GME approach, when the resources are scattered at various locations of area. To the best of our knowledge, the algorithm proposed in this paper is the first token-based GLME algorithm for MANETs. Algorithm satisfies the safety condition, starvation freedom and concurrent occupancy. Algorithm also handles link breakage and related dynamic changes in MANETs. To develop a fault tolerant version of the proposed algorithm can be a possible future research direction.

References

1. Swaroop, A.: Efficient Group Mutual Exclusion Protocols for Message Passing Distributed Computing System. Doctoral dissertation, NIT, Kurukshetra, India (2009)
2. Dijkstra, E.W.: Hierarchical ordering of sequential processes. Acta Informatica. **1**(2), 115–138 (1971)
3. Kshemkalyani, A.D., Singhal, M.: Distributed Computing Principles, Algorithms, Systems. Cambridge University Press, Cambridge (2009)
4. Joung, Y.J.: Asynchronous group mutual exclusion. Distrib. Comput. **13**(4), 189–206 (2000)
5. Joung, Y.J.: The congenial talking philosopher's problem in computer networks. Distrib. Comput. **15**(3), 155–175 (2002)
6. Attiya, H., Kogan, A., Welch, J.L.: Efficient and robust local mutual exclusion in mobile ad hoc networks. IEEE Trans. Mob. Comput. **9**(3), 361–375 (2010)
7. Kogan, A.: Efficient and Robust Local Mutual Exclusion in Mobile Adhoc Networks. Masters Research thesis, Technion—Israel Institute of Technology, Israel (2008)
8. Khanna, A., Singh, A.K., Swaroop, A.: A leader-based *k*-local mutual exclusion algorithm using token for MANETs. J. Inf. Eng. (JISE) (To appear) Taiwan (2014)
9. Luo, A., Wu, W., Cao, J., Raynal, M.: A generalized mutual exclusion problem and its algorithm. 42nd International Conference on Parallel Processing, pp. 300–309 (2013)

10. Sharma, B., Bhatia, R.S., Singh. A. K.: DMX in MANETs: major research trends since 2004. In: Proceedings of the International Conference on Advances In Computing and Artificial Intelligence ACAI'11, pp. 50–55 (2011)
11. Hosseinabadi, G., Vaidya, N.H.: Exploiting opportunistic overhearing to improve performance of mutual exclusion in wireless ad hoc networks. In: Proceeding of International Conference on Wired/Wireless Internet Communications (WWIC), pp. 162–173 (2012)
12. Parameswaran, M., Hota, C.: Arbitration-based reliable distributed mutual exclusion for mobile ad-hoc networks. Proc. Model. Optim. Mob. Ad Hoc Wirel. Netw. **11**, 380–387 (2013)
13. Wu, W., Cao, J., Yang, J.: Fault tolerant mutual exclusion algorithm for mobile ad hoc networks. Pervasive Mob. Comput. **4**, 139–160 (2008)
14. Tamhane, S.A., Kumar, M.: A token based distributed algorithm for supporting mutual exclusion in opportunistic networks. Pervasive Mob. Comput. **8**(5), 795–809 (2012)

Chapter 7
A Dual-band Z-shape Stepped Dielectric Resonator Antenna for Millimeter-wave Applications

Ashok Babu Chatla, Sanmoy Bandyopadhyay and B. Maji

Abstract Dual-band antennas are important in communication systems which operate in more than one frequency band. In this paper, a dual-band z-shaped stepped dielectric resonator antenna (DRA) is presented for millimeter-wave (MMW) applications. There are a number of definitions describing antennas operating in the MMW bands. Most commonly, antennas operating in frequencies whose wavelengths are measured in millimeters (30–300 GHz) are referred to as MMW antennas. Here, a novel three-step z-shaped dielectric resonator antenna is excited by a simple microstrip line feeding mechanism. Simulation process was done by using a CST microwave studio suite 2010. The result shows that the proposed antenna achieves an impedance bandwidth from 57.14 to 61.04 GHz and 64.06 to 69.01 GHz covering 58.77 GHz (unlicensed) band allocated for the earth exploration-satellite (EES) (passive) and fixed mobile space research (passive) services radio astronomy and 66.07 GHz band also allocated for EES and fixed inter-satellite mobile except aeronautical mobile space research on a primary basis, but in this band, both active and passive operations are permitted. The proposed design can also be used for inter-satellite service applications which operate at 65–66 GHz. With these features, this design of z-shaped stepped DRA is suitable for dual-band wireless communication systems. The federal communications commission has allocated the frequency band from 57 to 64 GHz for unlicensed wireless systems. This system is primarily used for high-capacity, short-distance (less than 1 mile) communications by any one.

A.B. Chatla (✉) · S. Bandyopadhyay · B. Maji
Department of Electronics and Communication Engineering,
National Institute of Technology, Durgapur, India
e-mail: ashokbabu23@gmail.com

S. Bandyopadhyay
e-mail: sanmoy1985@rediffmail.com

B. Maji
e-mail: bmajiecenit@yahoo.com

S. Patnaik et al. (eds.), *Recent Development in Wireless Sensor and Ad-hoc Networks*,
Signals and Communication Technology, DOI 10.1007/978-81-322-2129-6_7

Keywords Stepped DRA · A simple microstrip line feed · Earth exploration-satellite and space research services · Inter-satellite services · Millimeter-wave applications (57–64 GHz unlicensed)

7.1 Introduction

Satellite communication and wireless communication have been developed rapidly in the past decades, and it has already a dramatic impact on human life. In the last few years, the development of wireless local area networks (WLAN) represented one of the principal interests in the information and communication field. The field of wireless communications has been undergoing a revolutionary growth in the last decade. This is attributed to the invention of portable mobile phones some 15 years ago. The success of the second-generation (2G) cellular communication services motivates the development of wideband third generation (3G) cellular phones and other wireless products and services, including wireless local area networks, home RF, Bluetooth, wireless local loops, local multipoint distributed networks (LMDS), to name a few. The crucial component of a wireless network or device is the antenna. Very soon, our cities will be flooded with antennas of different kinds and shapes. On the other hand, for safety and portability reasons, low-power, multi-functional and multiband wireless devices are highly preferable. All these stringent requirements demand the development of highly efficient, low-profile and small-size antennas that can be made embedded into wireless products.

The increasing use of microwave mobile communication systems demands the antennas for different systems and standards with properties such as reduced size, broadband, multiband operation, and moderate gain. The planar and dielectric resonator antennas are the present-day antenna designer's choice. In the mid-1960s, Cohn and his coworkers at Rantec Corporation performed the first extensive theoretical and experimental evaluation of DR. Dielectric resonators (DRs), made of low-loss dielectric materials, are widely used in various applications. They are compact in size, are low in cost and weight, and could be excited using different transmission lines. Dielectric materials with dielectric constant values in the range of 10–100 have been used in applications at microwave frequencies. The high relative permittivity at the interfaces of DR and free space provides a standing electromagnetic wave inside the resonator. Besides the high dielectric constant and low loss, the stable temperature coefficient of these advanced materials offer new design opportunities for microwave applications. DRAs offer simple coupling schemes to nearly all transmission lines used at microwave and millimeter-wave frequencies.

This makes them suitable for integration into different planar technologies. The coupling between a DRA and the planar transmission line can be easily controlled by varying the position of the cylindrical dielectric resonator antenna (DRA) with respect to the line. The performance of DRA can therefore be easily optimized experimentally. The operating bandwidth of a DRA can be varied over a wide range by suitably choosing resonator parameters. For example, the bandwidth of the lower-order modes of a DRA can be easily varied from a fraction of a percent to about 10 % or more by the suitable choice of the dielectric constant of the resonator material.

DRAs offer advantages like many of the existing feeding schemes can be used (slots, probes, microstrip, coplanar waveguides, dielectric image guide, etc.) [1–4]. This makes them easy to integrate with existing technologies. Compared with the microstrip antenna, DRA has a much wider impedance bandwidth. This is because the microstrip antenna radiates only through two narrow radiation slots, whereas the DRA radiates through the whole antenna surface except the grounded part. Moreover, the operating bandwidth of a DRA can be varied by suitably choosing the dielectric constant of the resonator material and its dimensions.

DRs of different shapes have various modes of oscillation. With the proper excitation of certain modes and with no shielding, these resonators can actually become efficient radiators instead of energy storage devices. This concept led to the exploration of DRs as antennas. Although open DRs were found to radiate many years ago, the idea of using DR as a practical antenna materialized in 1983, when Long et al. published a paper on cylindrical DRAs. Subsequently, other researchers investigated rectangular and hemispherical DRAs. These works laid the foundation for later extensive investigations on various aspects of DRAs. Considerable work done on analyzing their resonant modes, radiation characteristics, and various feeding schemes has demonstrated that these “new” radiators could offer new and attractive features in antenna design.

7.1.1 Dielectric Resonators

A dielectric resonator (dielectric resonator oscillator, DRO) is an electronic component that exhibits resonance for a narrow range of frequencies, generally in the microwave band. The resonance is similar to that of a circular hollow metallic waveguide except that the boundary is defined by large change in permittivity rather than by a conductor. Dielectric resonators generally consist of a “puck” of ceramic that has a large dielectric constant and a low dissipation factor. The resonance frequency is determined by the overall physical dimensions of the puck and the dielectric constant of the material.

Although dielectric resonators display many similarities to resonant metal cavities, there is one important difference between the two: while the electric and magnetic fields are zero outside the walls of the metal cavity (i.e., open circuit boundary conditions are fully satisfied), these fields are not zero outside the dielectric walls of the resonator (i.e., open circuit boundary conditions are approximately satisfied). Even so, electric and magnetic fields decay from their maximum values considerably when they are away from the resonator walls. Most of the energy is stored in the resonator at a given resonant frequency for a sufficiently high dielectric constant ε_r. Dielectric resonator can exhibit extremely high *Q*-factor that is comparable to a metal-walled cavity.

7.1.2 Common Applications of DRO

Most common applications of dielectric resonators are as follows:

- Filtering applications (most common are band-pass and band-stop filters),
- Oscillators (diode, feedback-, reflection-, transmission-, and reaction-type oscillators),
- Frequency-selective limiters,
- DRA elements.

7.1.3 Historical Overview

In the late nineteenth century, Lord Rayleigh demonstrated that an infinitely long cylindrical rod made up of dielectric material could serve as a waveguide. Additional theoretical and experimental work done in Germany in early twentieth century offered further insight into the behavior of electromagnetic waves in dielectric rod waveguides. Since a dielectric resonator can be thought of as a truncated dielectric rod waveguide, this research was essential for scientific understanding of electromagnetic phenomena in dielectric resonators. In 1939, Robert D. Richtmyer published a study [5] in which he showed that dielectric structures can act just as metallic cavity resonators. He appropriately named these structures dielectric resonators. Richtmyer also demonstrated that, if exposed to free space, dialectic resonators must radiate because of the boundary conditions at the dielectric-to-air interface. These results were later used in development of DRA. Due to World War II, lack of advanced materials and adequate manufacturing techniques, dielectric resonators fell in relative obscurity for another two decades after Richtmyer's study was published.

However, in 1960s, as high-frequency electronics and modern communications industry started to take off, dielectric resonators gained in significance. They offered a size-reducing design alternative to bulky waveguide filters and lower-cost alternatives for electronic oscillator, frequency-selective limiter, and slow-wave circuits. In addition to cost and size, other advantages that dielectric resonators have over conventional metal cavity resonators are lower weight, material availability, and ease of manufacturing.

7.1.4 Investigations on Dielectric Resonator Antenna

For many years, the DR has primarily been used in microwave circuits, such as oscillators and filters, where the DR is normally made of high-permittivity material, with dielectric constant $\varepsilon_r > 20$. The unloaded Q-factor is usually between 50 and 500, but can be as high as 10,000. Because of these traditional applications, the DR was usually treated as an energy storage device rather than as a radiator. Although open DRs were found to radiate many years ago, the idea of using the DR as an antenna had not been widely accepted until the original paper on the cylindrical DRA [6] was published in 1983. At that time, it was observed that the frequency range of interest for many systems had gradually progressed upward to the millimeter and near-millimeter range (100–300 GHz). At these frequencies, the conductor loss of metallic antennas becomes severe and the efficiency of the antennas is reduced significantly. Conversely, the only loss for a DRA is that due to the imperfect dielectric material, which can be very small in practice. After the cylindrical DRA had been studied, Long and his colleagues subsequently investigated the rectangular and hemispherical DRAs. The work created the foundation for future investigations of the DRA. Other shapes were also studied, including the spherical-cap [7], and cylindrical-ring [8], triangular DRAs [9]. Figure 7.1 shows a photograph of various DRAs [2, 10]. It was found that DRAs operating at their fundamental modes radiate like a magnetic dipole, independent of their shapes.

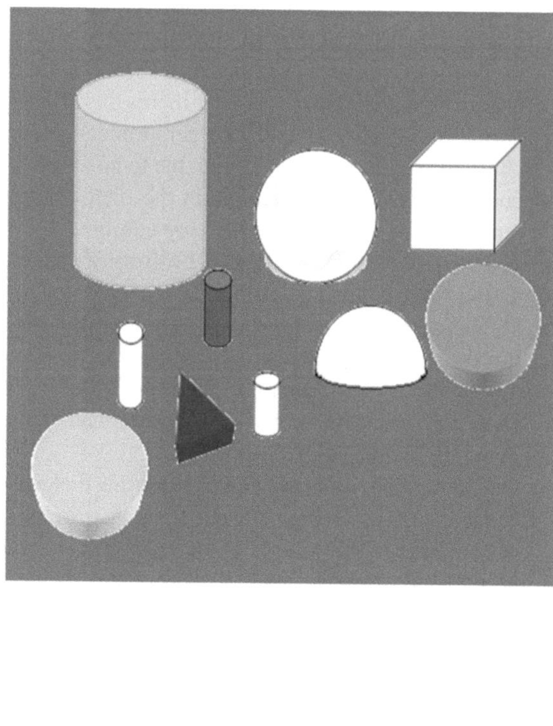

Fig. 7.1 DRAs of various shapes, *cylindrical*, *rectangular*, *hemispherical*, *low-profile circular-disk*, and *low-profile triangular*, and *spherical-cap* DRA

7.1.5 Advantages Over Microstrip Antennas

As compared to the microstrip antenna, the DRA has a much wider impedance bandwidth ($\sim$10 % for dielectric constant $\sim$10). This is because the microstrip antenna radiates only through two narrow radiation slots, whereas the DRA radiates through the whole DRA surface except the grounded part [11]. Avoidance of surface waves is another attractive advantage of the DRA over the microstrip antenna. Nevertheless, many characteristics of the DRA and microstrip antenna are common because both of them behave like resonant cavities. For example, since the dielectric wavelength is smaller than the free-space wavelength by a factor of $1/\sqrt{\varepsilon_r}$, both of these can be made smaller in size by increasing ε_r.

7.1.6 Characteristics of the Dielectric Resonator Antennas

For different applications, the dielectric resonator has been studied extensively in recent years from low frequencies to high frequencies. The dielectric resonator antenna has versatility and high flexibility in adopting a shape over a wide frequency range, which allows the desired requirement to the designer. Some of the major advantages or characteristics of the dielectric resonator antenna are as follows:

- The size of the dielectric resonator antenna is proportional to $\sqrt{\varepsilon_r}$, where is the free-space wavelength of the resonant frequency and ε_r is the dielectric constant of the dielectric resonator. This represents that the size of the dielectric resonator antenna can be reduced by just increasing the dielectric constant [10].
- A millimeter-wave frequency operation can be achieved by choosing a low-loss characteristic dielectric material due to an absence of surface waves and minimal conductor losses associated with the dielectric resonator antenna. At these frequencies, high radiation efficiency can be achieved [11].
- The dielectric constant ε_r can be from a range of below 3 up to 100, which allows for great deal of flexibility when controlling the size and bandwidth [12].
- Different methods can be used to excite the dielectric resonator antenna (slot, probe, coplanar, microstrip, waveguide, dielectric image guide, etc.) which make them easy to integrate with the existing technologies [10].
- Dielectric resonator antennas are designed to operate over a wide frequency range of 1.3–40 GHz [11].
- Dielectric resonator antennas have high dielectric strength which makes them capable to handle high power. This also helps them to work in a wide temperature range due to the temperature-stable ceramic materials.
- Many modes can be excited within the dielectric resonator antenna element depending on the shape of the resonator. Different modes give different radiation patterns for various coverage requirements.

- The radiation Q-factor of the modes depends on the aspect ratio; the aspect ratio is an important parameter in designing a dielectric resonator antenna because it gives one more degree of freedom for the design [11].
- The dielectric resonator antenna has much wider impedance bandwidth compared to the microstrip antenna. It is because the dielectric resonator antenna radiates through the whole antenna surface except for the ground, whereas the microstrip antenna radiates only through the narrow slots [12].

The fabrication of dielectric resonator antennas is more complex and more costly compared to printed circuit antennas, especially for array applications. Maybe for high-volume of production the cost reduces but the economic scale is not that important. The problem with dielectric resonator antennas is typically that ceramic materials which are machined to shape from a large block are used. To feed the antenna excitations from, e.g., probes, the antenna needs to be drilled since it is bounded to a ground plane or a substrate. There are applications where performance is more important than cost and the dielectric resonator antenna can provide the solution.

7.1.7 Excitation Methods Applied to the DRA

A number of excitation methods have been developed [1–4]. Examples are the coaxial probe, aperture coupling with a microstrip feed line, aperture coupling with a coaxial feed line, direct microstrip feed line, co-planar feed, soldered-through probe, slot line, strip line conformal strip, and dielectric image guide is shown in Fig. 7.3 [13]. A photograph of the coaxial probe excitation scheme is shown in Fig. 7.2 [2, 10]. Some of the feeding methods are addressed, whereas the rigorous analyses of the aperture coupling with a perpendicular feed and conformal strip feed are presented. Virtually all excitation methods applicable to the microstrip antenna can be used for the DRA.

Fig. 7.2 A probe-fed DRA. Above the ground plane is the coaxial probe

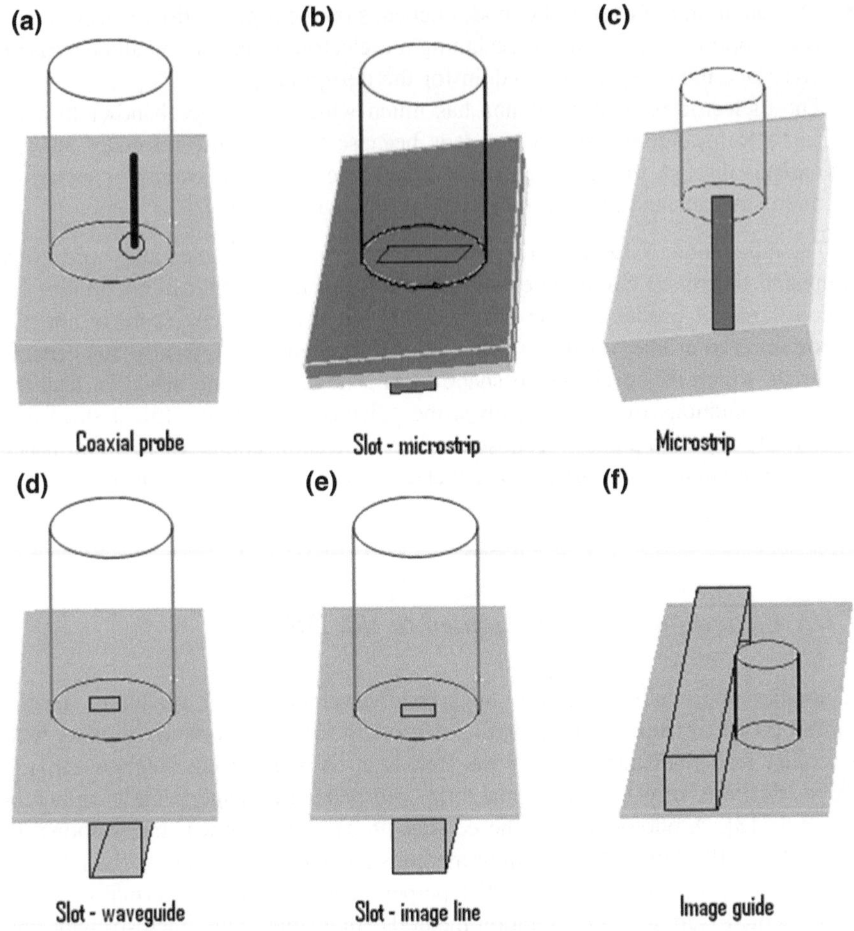

Fig. 7.3 Different excitation mechanisms for the DRA

7.1.8 Materials for Dielectric Resonators

Different materials have different applications and requirements of the dielectric resonator. The dielectric properties depend on the crystal structure, material, and imperfections in the crystal lattice as well as on the preparation conditions. Dielectric resonators generally consist of a puck-formed cylinder of ceramic material of high permittivity and low dissipation factor. Traditional passive devices have been desired to have a high Q-factor up to 20,000 between 2 and 20 GHz [12] high permittivity and near-zero temperature coefficients for the resonant frequency which has been difficult to achieve simultaneously. The low-permittivity ceramics have so far been used for millimeter-wave applications and also as substrates for microwave integrated circuits. Ceramics with permittivity in the range between 25 and 50 have been used for satellite communication and in cell phone base stations. High-permittivity materials are used in mobile telephones, where the compact size sets narrow limits. In all the mentioned cases, the use has been for circuit components. Materials of lower relative permittivity are Teflon-based plastic materials. One provider of these materials is the Rogers Corporation, with sheets of materials with permittivity from around 2 to 10.2.

7.1.9 Millimeter-wave Overview

Millimeter wave generally corresponds to the radio spectrum between 30 and 300 GHz, with wavelength between one and ten millimeters. However, in the context of wireless communication, the term generally corresponds to a few bands of spectrum near 38, 60 and 94 GHz, and more recently to a band between 70 and 90 GHz (also referred to as E-band), that have been allocated for the purpose of wireless communication in the public domain.

Though relatively new in the world of wireless communication, the history of millimeter-wave technology goes back to the 1890s when J.C. Bose was experimenting with millimeter-wave signals at just about the time when his contemporaries like Marconi were inventing radio communications. Following Bose's research, millimeter-wave technology remained within the confines of university and government laboratories for almost half a century. The technology started so see its early applications in radio astronomy in the 1960s, followed by applications in the military in the 1970s. In the 1980s, the development of millimeter-wave integrated circuits created opportunities for mass manufacturing of millimeter-wave products for commercial applications [14].

In the 1990s, the advent of automotive collision avoidance radar at 77 GHz marked the first consumer-oriented use of millimeter-wave frequencies above 40 GHz. In 1995, in the FCC US Federal Communications in the USA, four bands in the upper millimeter-wave region have been opened for commercial applications. Of the four bands, the 59–64 GHz band commonly referred to as V-band or the 60 GHz band is governed by FCC Part 15 for unlicensed operations. The regulations of FCC Part 15 and the significant absorption of the 60 GHz band by atmospheric oxygen make this band better suited for very-short-range point-to-point and point-to-multipoint applications. The 92–95 GHz band commonly referred to as W-band or the 94 GHz band is also governed by the FCC Part 15 regulations for unlicensed operation, though for indoor applications only. The 94 GHz band may also be used for licensed outdoor applications for point-to-point wireless communication per FCC Part 101 regulations. However, the band is less spectrally efficient than the other three bands due to an excluded band at 94–94.1 GHz.

This leaves 71–76 GHz and 81–86 GHz commonly referred to as E-band or the 70 and 80 GHz bands, respectively, whose use in the USA is governed by FCC Part 101 for licensed operation, as the most ideally suited millimeter-wave band for point-to-point wireless communication applications. With the 5 GHz of spectrum available in each of these two bands, the total spectral bandwidth available exceeds that of all allocated bands in the microwave spectrum. Through the remainder of this paper, we confine our discussion to these two bands.

The 5 GHz of spectrum available in each sub-band of the E-band spectrum can be used as a single, contiguous transmission channel which means no channelization is required, thereby allowing the most efficient use of the entire band. Even with simple modulation techniques such as on-off-keying (OOK) or binary phase shift keying (BPSK), throughput of 1–3 Gbps is achieved today in each sub-band of the spectrum, more that what can be achieved with more sophisticated, higher-order modulation schemes in other bands of licensed spectrum. Migrating to such modulation techniques at E-band, even higher throughputs can be achieved. It is only a matter of sufficient market demand before such higher throughput millimeter-wave links become a commercial reality.

In the past few years, the interest in the millimeter-wave spectrum at 30–300 GHz has drastically increased. The emergence of low-cost high-performance CMOS technology and low-loss, low-cost organic packaging material has opened a new perspective for system designers and service providers because it enables the development of millimeter-wave radio at the same cost structure of radios operating in the gigahertz range or less. In combination with available ultra-wide bandwidths, this makes the millimeter-wave spectrum more attractive than ever before for supporting a new class of systems and applications ranging from ultra-high-speed data transmission, video distribution, portable radar, sensing, detection, and imaging of all kinds.

While at a lower frequency, the signal can propagate easily for dozens of kilometers, penetrate through construction materials, or benefit from advantageous reflection and refraction properties, one must consider carefully the characteristics in particular strong attenuation and weak diffraction of the millimeter-wave propagation, and exploit them advantageously. The free-space loss (FSL) after converting to units of frequency and putting them in decibel form between two isotropic antennas can be expressed as [15]. As an example, at 60 GHz the FSL is much more severe than at the frequencies usually used for cell phone and wireless applications. The link budget at 60 GHz is 21 dB less than the one at 5 GHz under equal conditions. In addition, other loss and fading factors increasingly affect the millimeter-wave transmission, such as gasses, rain, foliage, and scattering and diffraction losses.

7.1.10 Key Benefits of Millimeter Wave

The key benefits of millimeter wave can be summarized as

- Ultra-high-speed data transmission
- Unmatched bandwidth with scalable capacity
- Narrow beam with highly scalable deployments
- Licensed spectrum with low-cost licensing
- Mature technology with multivendor solutions
- Detection and imaging of all kinds

7.1.11 Application Areas for Millimeter Wave

- Metro network services:

With the economy becoming more information dependent, the bandwidth needs of corporations, large and small, continue to grow apparently without bound. However, a large majority of corporate buildings are still being served only by archaic copper wires barely able to deliver a few megabits per second of bandwidth.

- Cellular/WiMAX backhaul:

With the use of mobile handheld devices growing and newer bandwidth-intensive applications emerging, the need to deliver higher bandwidth to mobile users will continue to rise. As newer technologies such as WiMAX and new spectrum such as 700 MHz are used to serve these needs at the access point, the need for a technology to transport the bandwidth from the point of access to the core of the network will rise swiftly. To this day, most of those needs have been met by slower capacity channels such as T1/E1 leased lines. However, these solutions will not be able to meet the needs of the next generation of mobile networks in a practical manner.

- Cellular distributed antenna systems (DAS):

In cellular networks, it is often necessary or more efficient to enhance network coverage by distributing a network of remote antennas instead of providing coverage by way of centrally located antennas. Such DAS are basically extensions of the antenna of base stations. DAS are often used to provide cellular coverage in spots that are shadowed by large structures, such as buildings, from base station antennas. DAS may also be used to provide coverage in areas, where it is not efficient to install a base station.

- Failure recovery and redundancy:

For applications requiring high end-to-end bandwidth, broadband connectivity by means of fiber optic cables is often the technology of choice when access to fiber optic cables is readily available. However, cases abound where fiber connections have been broken by accident, for instance during trenching operations, often bringing down mission critical networks for a substantial period of time. Therefore, it is highly desirable to design such mission critical networks with redundancies that minimize probability of such failures.

- Enterprise and campus networks:

The needs of enterprises to extend LANs from one building to a neighboring building are often so compelling that users in such applications have been the earliest adopter of point-to-point wireless technologies. As organizations expand their facilities by growing into neighboring buildings, the cost of leasing interconnecting communication services becomes significant, eventually persuading them to look for alternate solutions.

This paper introduces a novel z-shaped dielectric resonator antenna which can provide a large bandwidth. As shown in Fig. 7.1, it has a stepped structure and can be envisaged as dielectric bricks of unequal size fused together. Due to stepped structure of the antenna, its effective aperture is expected to be larger than that of conventional DR antennas [4].

The antenna has a ground plane and is fed through a simple microstrip line. To show the potentials of the antenna, for a particular three-step DR antenna using a material of dielectric constant $\varepsilon_r = 10.2$, simulation results for the return-loss and radiation patterns over different frequencies within the impedance bandwidth are presented and discussed in the result section. However, earth exploration-satellite (EES) passive and fixed mobile space research passive radio astronomy (58.2–59 GHz) and EES and fixed inter-satellite mobile except aeronautical mobile space research on a primary basis but in this band active and passive operations are permitted and this band is allocated for the inter-satellite service (65–66 GHz) millimeter-wave applications are intended this proposed novel z-shaped stepped DRA [16].

7.2 Motivation

The main motivation of this work is nowadays, mobile communication system is becoming increasingly popular. Antennas for software-defined and/or reconfigurable radio systems need to have ultra-wide-band or multiband characteristics in order to be flexible enough to cover any possible future mobile communication frequency bands. One approach to provide such flexibility is to construct multiband antenna that operates over specific narrowband frequencies.

However, it would be extremely difficult to accurately achieve the frequency requirements of all future communication system. Alternatively, a small wideband antenna that covers a wide range of frequencies can be a good candidate not only for current multiband applications but also for future communication systems operating on new frequency bands. With bandwidths as low as a few percent, wideband applications using conventional microstrip patch designs and DRA designs are limited. Other drawbacks of patch antennas include low efficiency, limited power capacity, spurious feed radiation, poor polarization purity, narrow bandwidth, and manufacturing tolerance problems.

To overcome these problems for over two decades, research scientists have developed several methods to increase the bandwidth and low-frequency ratio of a patch antenna and DRA. The motivation to extend dielectric resonator antenna for millimeter applications due to the dimension of a DRA is of the order of $/\sqrt{\varepsilon r}$, where ε_r is the dielectric constant of the resonator material and is the free-space wavelength. Thus by choosing a high value of ε_r (10–100), the size of the DRA can be significantly reduced. There is no inherent conductor loss in dielectric resonators. This leads to high radiation efficiency of the antenna. This feature is especially attractive for millimeter (mm)-wave antennas, where the loss in metal-fabricated antennas can be high. Many of these techniques involve adjusting the placement and/or type of element used to feed (or excite) the antenna. Wide-band frequency operation of antennas has become a necessity for many applications. Microstrip antennas and DRAs are ultimately expected to replace conventional antennas for most applications.

- Requirements for the digital home include high-speed data transfer for multimedia content, short-range connectivity for transfer to other devices
- Satellite communication system
- Radar and imaging
- Medical applications
- Location and tracking
- Vehicular radar systems
- Low-data-rate and low-power UWB communications

The basic applications of DRAs lie in millimeter-wave applications. Extremely high frequency is the highest radio frequency band. EHF runs the range of frequencies from 30 to 300 GHz, above which electromagnetic radiation is considered to be low (or far) infrared light, also referred to as terahertz radiation. This band has a wavelength of 10–1 mm, giving it the name millimeter band or millimeter wave, sometimes abbreviated MMW or mmw. One use for the millimeter wave is in the transmission of a great deal of data. Millimeter waves are used in radio broadcasting, cell phone transmissions, and satellites. There is a huge difference in the amount of data sent during a phone call versus a television program. This is what makes the millimeter wave so important.

7.3 Antenna Design

Figure 7.4 shows the geometry of the proposed DRA, where z-shaped stepped dielectric resonators having dielectric constant 10.2 are placed above a substrate with a dielectric constant 2.2. Below the substrate is a ground plane. The dimension of the ground plane is 10×12 mm^2 ($l \times w$) and height is $h_g = 0.5$ mm. The same dimension ($l \times w$) is used for substrate also and height is $h_s = 1.6$ mm. The z-shaped stepped DRA consists of four dielectric bricks of unequal size fused together. One-, two-, and three-stepped DR antennas with a finite ground plane were analyzed using CST microwave studio software [17].

Due to space constraint, only the results for the three-step structure are given here. However, it is worth noting that the one-step structure is essentially the same as the rectangular DR antenna reported in [18]. In fact, the structure was analyzed first in order to confirm the accuracy of the software used [18]. The physical and electrical properties of the dielectric structure used in the three-step DR antenna are shown in Fig. 7.4.

The width, length, and the height of each brick are 5.5, 4, 1.7 mm; 4, 2, 0.35 mm; 3, 2.5, 0.3 mm; and 2, 2, 1 mm, respectively. The simple microstrip feed line is etched on FR4 substrate with width Wf = 6 mm, length Lf = 0.5 mm, and height Hf = 0.5 mm. Here by varying the height of patch for coupling, the resonant

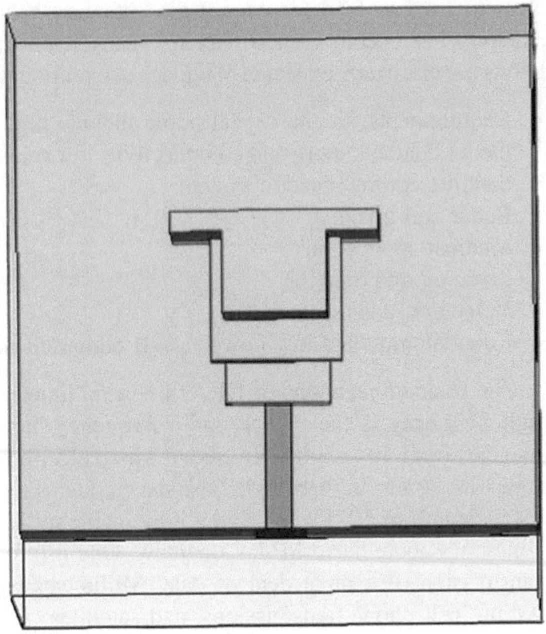

Fig. 7.4 Geometry (*top view*) of the proposed z-shaped stepped DRA. Here, *x*-axis is length, *y*-axis is width, and *z*-axis is height

frequency is also varying, whereas the depths of the steps are chosen by trial and error in order to obtain the dual bandwidth.

The dual-band design of the proposed z-shaped stepped DRA adopts different methods [16–19]. The coupling between the DR and the feed mechanism can be easily adjusted by changing the size of the conformal patch; thus, a dual-band impedance matching has been obtained. The desired frequencies for unlicensed EES passive and fixed mobile space research passive radio astronomy and EES and fixed inter-satellite mobile except aeronautical mobile space research on a primary basis but in this band active and passive operation are permitted and this band is allocated for the inter-satellite service millimeter-wave applications are obtained [14].

7.4 Simulation and Experimental Results

7.4.1 Return-loss Characteristics

A dual-band z-shaped stepped DRA for 58.77 and 66.07 GHz millimeter-wave applications has been designed and analyzed. Figure 7.5 shows the simulated S parameters. The proposed antenna is simulated and a return loss of −10 dB below dual band is obtained which is in the range of 57.14–61.04 GHz and 64.06–69.01 GHz. Here, a good return loss is obtained at two peaks whose values are −25.9 dBi at 58.77 GHz and a value of −21.5 dBi at 66.07 GHz.

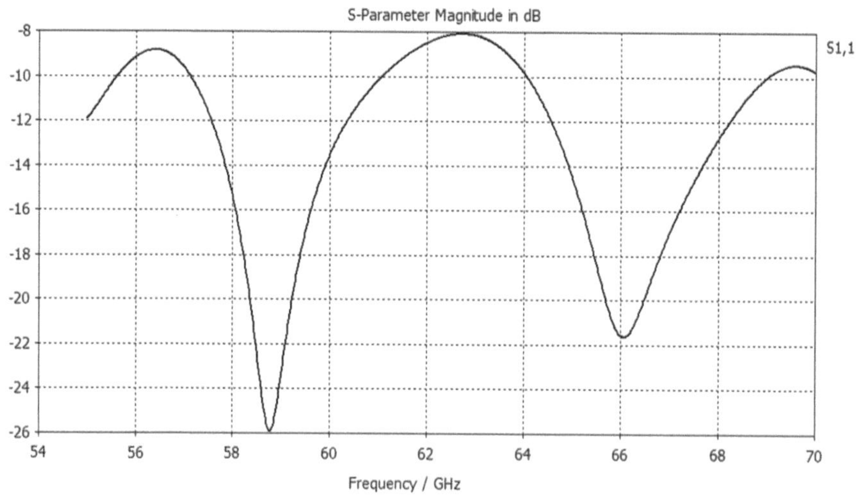

Fig. 7.5 Return loss versus frequency plot of proposed a dual-band z-shaped stepped DRA

7.4.2 Voltage Standing Wave Ratio

Figure 7.6 shows the voltage standing wave ratio (VSWR) is a well-known indication of how good the impedance match is. The VSWR is often abbreviated as SWR. A high VSWR is an indication that the signal is reflected prior to being radiated by the antenna. VSWR and reflected power are different ways of measuring and expressing the same thing. VSWR parameter of the proposed antenna is obtained below 2 dBi at 57.14–61.04 GHz and 64.06–69.01 GHz.

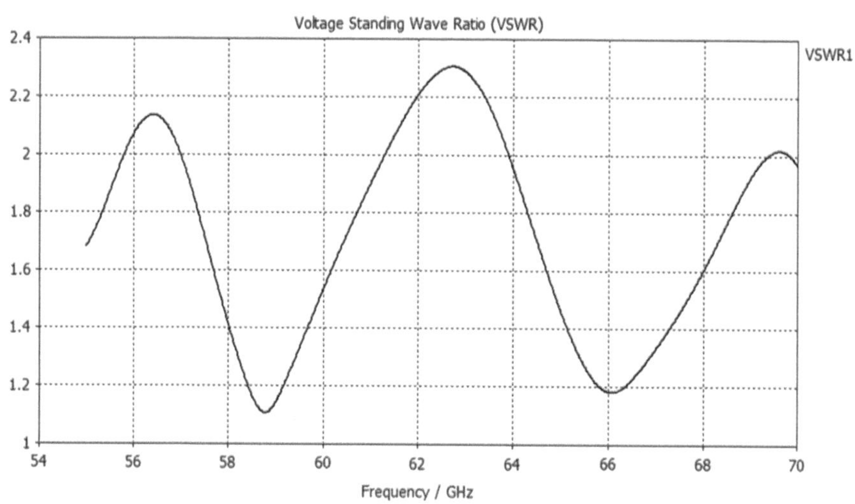

Fig. 7.6 VSWR versus frequency plot of proposed DRA

7.4.3 Surface Current Flows of Proposed DRA

The H-field peak surface current flow of the proposed DRA at 58.77 GHz frequency and 0° phase containing 110.105 A/m maximum—3D refer in Fig. 7.7a. The H-field peak surface current flow of the proposed DRA at 66.07 GHz frequency and 0° phase containing 122.246 A/m maximum—3D refer in Fig. 7.7b.

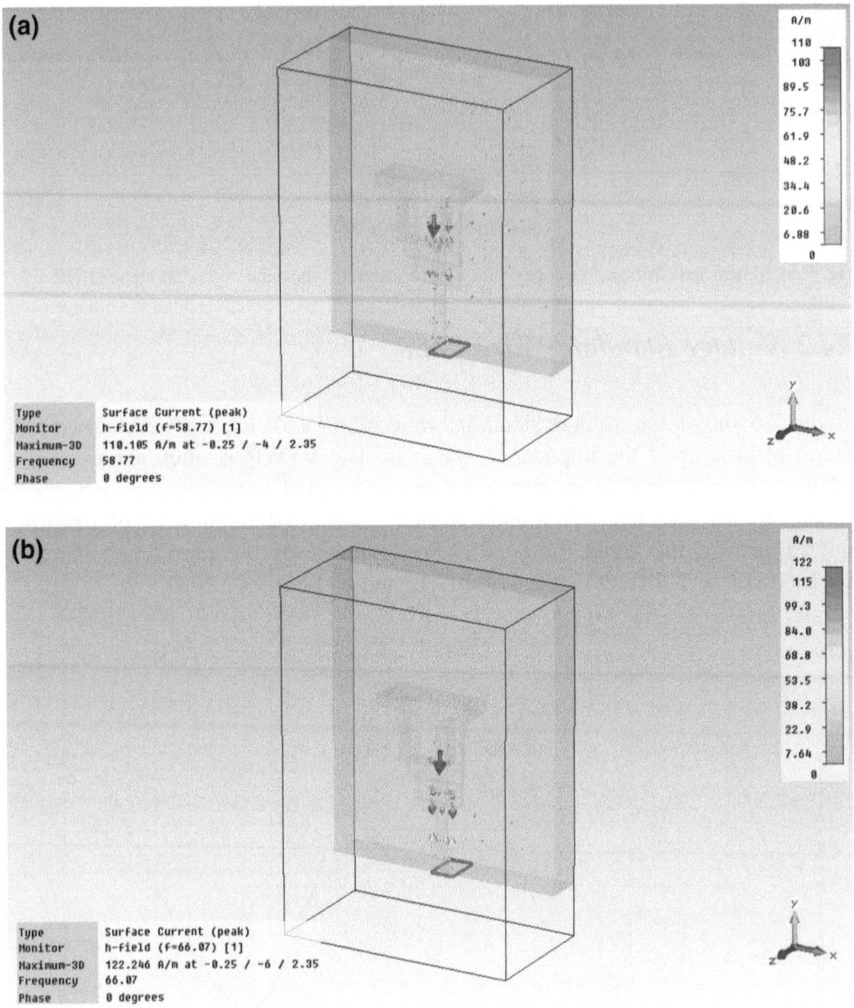

Fig. 7.7 **a** Surface current flow of the proposed antenna at frequency 58.77 GHz. **b** Surface current flow of the proposed antenna at frequency 66.07 GHz

7.4.4 3D View Radiation Patterns of Gain and Directivity

Figure 7.8a–d shows 3D plots of radiation patterns; here, directivity of 12.5 dBi at 58.77 GHz and a 12.1 dBi at 66.07 GHz has been claimed, besides that a significant gain of 11.91 dBi at a frequency of 58.77 GHz and 11.91 dBi at 66.07 GHz.

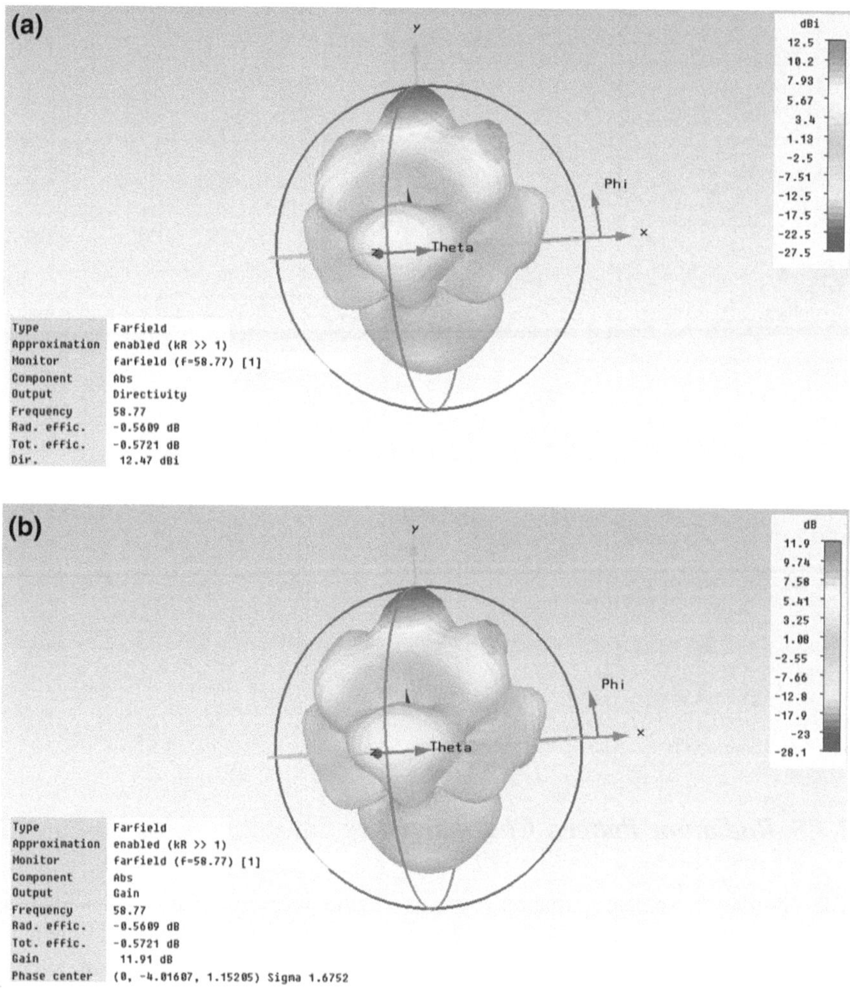

Fig. 7.8 **a** The directivity of proposed DRA at frequency 58.77 GHz. **b** The directivity of proposed DRA at frequency 66.07 GHz. **c** The gain of proposed DRA at frequency 58.77 GHz. **d** The gain of proposed DRA at frequency 66.07 GHz

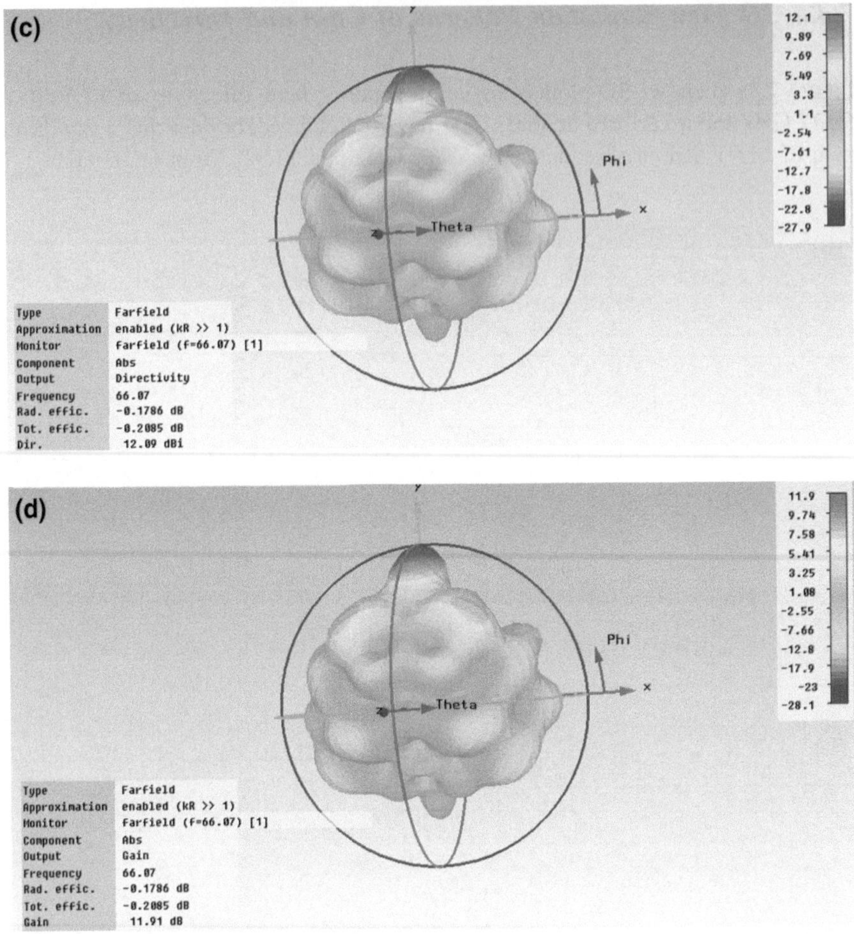

Fig. 7.8 (continued)

7.4.5 Radiation Pattern Characteristics

The simulated far-field radiation patterns of the proposed DRA are shown in Fig. 7.9a–d. It shows the simulated radiation patterns at different frequencies (58.77 GHz, 66.07 GHz). It has been observed that the *E*-plane radiation patterns and *H*-plane radiation patterns are different for all frequencies.

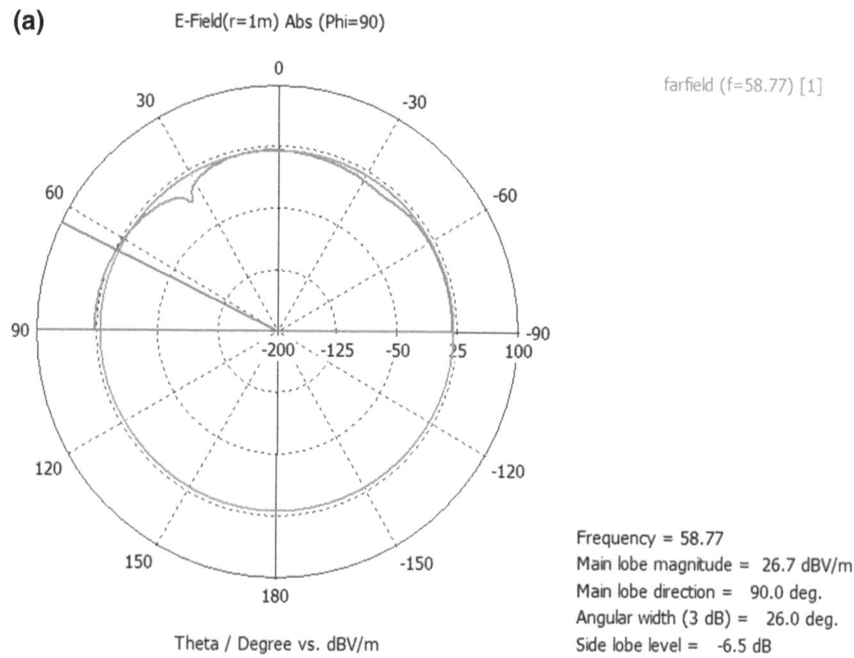

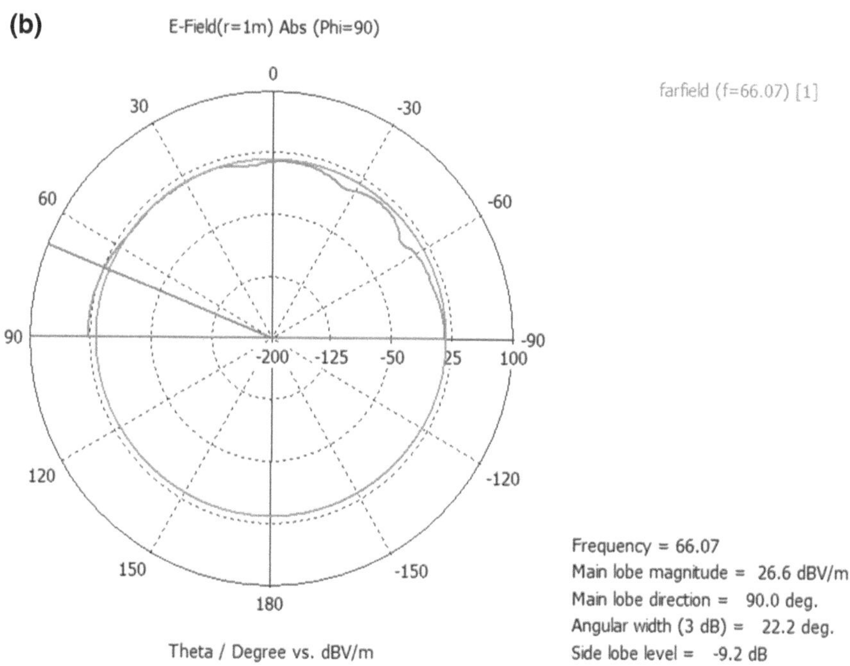

Fig. 7.9 **a** The E-field radiation pattern of proposed DRA at frequency 58.77 GHz. **b** The E-field radiation pattern of proposed DRA at frequency 66.07 GHz. **c** The H-field radiation pattern of proposed DRA at frequency 58.77 GHz. **d** The H-field radiation pattern of proposed DRA at frequency 66.07 GHz

(c)

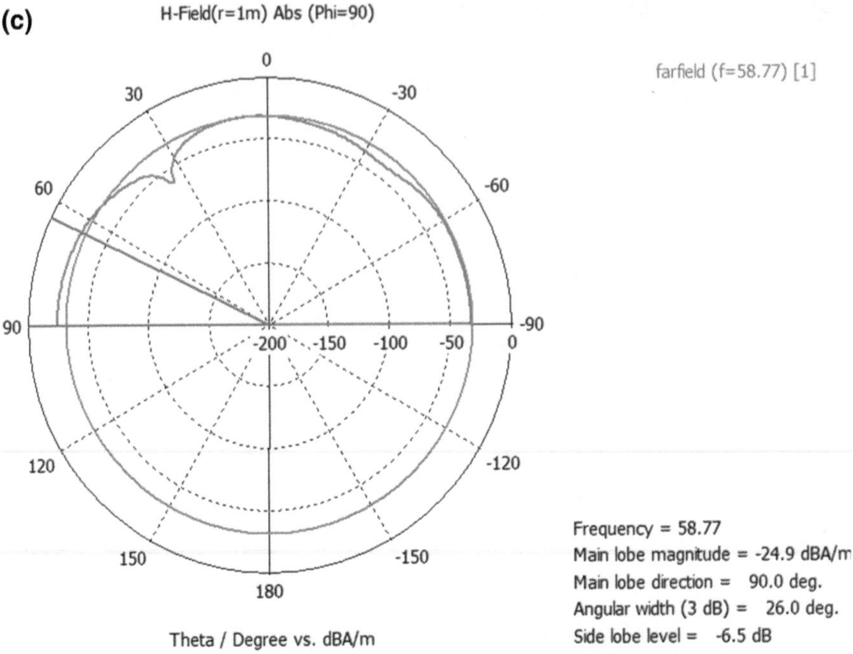

(d)

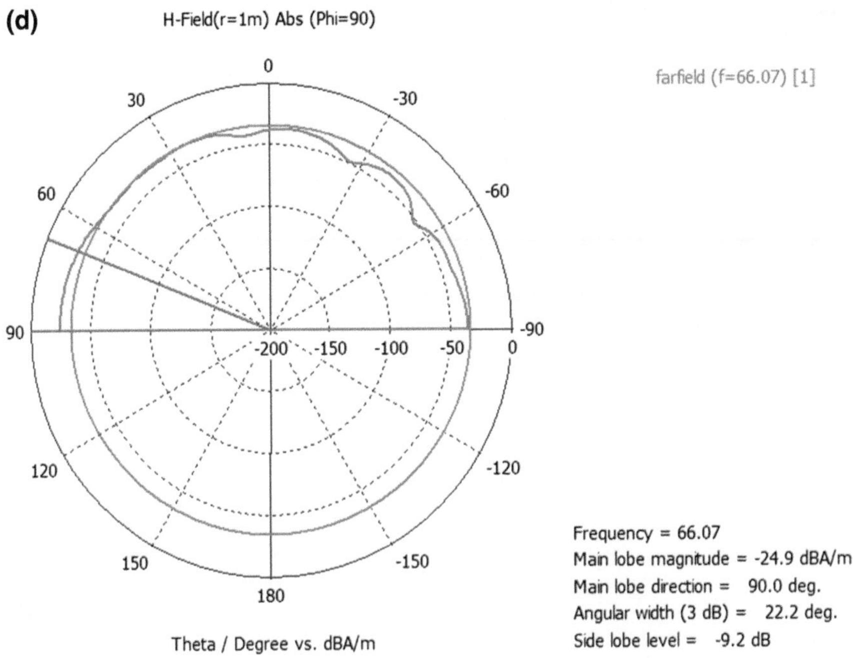

Fig. 7.9 (continued)

7.5 Conclusion

In this paper, a dual-band z-shaped stepped dielectric resonator antenna [20] is presented to obtain the desired resonant frequency for millimeter-wave applications. The z-shaped stepped DRA consists of four dielectric bricks of unequal size fused together. These bricks are excited by simple microstrip line. The simulated results show that the designed antenna offered impedance bandwidth from 57.14 to 61.04 GHz and 64.06 to 69.01 GHz, which covers several important application bands in current wireless communication systems. This antenna design provides a maximum gain of 11.91 dBi and directivity of 12.5 dBi. The presented dual-band antenna is suitable for EES passive and fixed mobile space research passive radio astronomy (58.2–59 GHz) and EES and fixed inter-satellite mobile except aeronautical mobile space research on a primary basis but in this band active and passive operations are permitted and this band is allocated for the inter-satellite service (65–66 GHz) millimeter-wave applications. The federal communications commission has allocated the frequency band from 57 to 64 GHz for unlicensed wireless systems. This system is primarily used for high-capacity, short-distance (less than 1 mile) communications by any one.

We have implemented this particular antenna using CST software. CST microwave studio is one of the most imperial electromagnetic software which allows to solving for radio and microwave application. CST MICROWAVE STUDIO® is a fully featured software package for electromagnetic analysis and design in the high-frequency range [21]. It simplifies the process of inputting the structure by providing a powerful solid modeling front-end which is based on the famous ACIS modeling kernel. Strong graphic feedback simplifies the definition of our device even further. After the component has been input, a fully automatic meshing procedure (based on an expert system) is applied before the simulation engine is started. The simulator itself features the new perfect boundary approximation (PBA method) and its thin-sheet technique (TST) extension, which increases the accuracy of the simulation by an order of magnitude in comparison with conventional simulators. Since no method works equally well in all application domains, the software contains four different simulation techniques (transient solver, frequency domain solver, eigenmode solver, and modal analysis solver) which best fit their particular applications. The "original" solver in CST Microwave Studio is based on finite integration technique. The simulator tool computes most of the useful quantities of interest such as radiation pattern, reflection coefficient, VSWR, and gain.

Acknowledgments For our work, we wish to acknowledge the Department of Electronics and Communication Engineering of National Institute of Technology, Durgapur, for helping us through out the project by giving us the good laboratory facility to continue our work in this field of a dual-band z-shaped stepped dielectric resonator antenna for millimeter-wave applications.

References

1. Petosa, A.: Dielectric Resonator Antenna Handbook. Artech House Publishers (2007)
2. Luk, K.M., Leung, K.W.: Dielectric Resonator Antennas. Research Studies Press Ltd, Hertfordshire (2002)
3. Mongia, R.K., Bhartia, P.: Dielectric resonator antennas—a review and general design relation for resonant frequency and bandwidth. Int. J. Microw. Millimeter-Wave Computer-Aided Eng. **4**(3), 230–247 (1994)
4. Petosa, A., Ittipiboon, A., Antar, Y.M.M., Roscoe, D., Cuhaci, M.: Recent advances in dielectric resonator antenna technology. IEEE Antennas Propag. Mag. **40**(3), 35–48 (1998)
5. Richtmyer, R.D.: Dielectric resonators. J. Appl. Phys. **10**, 391–398 (1939)
6. Long, S.A., McAllister, M.W., Shen, L.C.: The resonant cylindrical dielectric cavity antenna. IEEE Trans. Antennas Propag. **31**, 406–412 (1983)
7. Leung, K.W., Luk K.M., Yung E.K.N.: Spherical cap dielectric resonator antenna using aperture coupling. Electron. Lett. **30**(17), 1366–1367 (1994)
8. Mongia, R.K., Ittipiboon, A., Bhartia, P., Cuhaci, M.: Electric monopole antenna using a dielectric ring resonator. Electron. Lett. **29**, 1530–1531 (1993)
9. Lo, H.Y., Leung, K.W., Luk K.M., Yung, E.K.N.: Low-profile equilateral triangular dielectric resonator antenna of very high permittivity. Microwave Opt. Lett., **35**(25), 2164–2165 (2000)
10. Luk, K.M., Leung, K.W.: Dielectric Resonator Antennas. Research Studies Press Ltd, Baldock (2003)
11. A. Petosa: Dielectric Resonator Antenna Handbook, Artech House Publishers (2007)
12. Kishk, A.A., Antar, Y.M.M.: Dielectric resonator antennas, from J.L. Volakis: Antenna Engineering Handbook, Chapter 17, 4th edn., McGraw-Hill, USA (2007)
13. Volakis, J.L.: Antenna Engineering Hand Book, 4th edn
14. Alwis, G.: Murray Delahoy spectrum planning and Engineering Team Radio frequency planning Group Australian Communications Authority. 60 GHz Band Millimeter Wave Technology, Document: 3/04, Date: December 2004
15. Mongia, R.K., Ittipiboon, A.: Theoretical and experimental investigations on rectangular dielectric resonator antennas. IEEE Trans. Antennas Propag. **45**, 1348–1356 (1997)
16. Buerkle, A., Sarabandi, K., Mosallaei, H.: Compact slot and dielectric resonator antenna with dual-resonance, broadband characteristics. IEEE Trans. Antennas Propag. **53**(3), 1020–1027 (2005)
17. Denidni, T.A., Rao, Q., Sebak, A.R.: Multi-eccentric ring slot-fed dielectric resonator antennas for multi-frequency operations. IEEE Int. Symp. Antennas Propag. **2**, 1379–1382 (2004)
18. Denidni, T.A., Rao, Q.: Hybrid dielectric resonator antennas with radiating slot for dual-frequency operation. IEEE Antennas Wireless Propag. Lett. **3**(1), 321–323 (2004)
19. Li, T.W., Sun, J.S.: Dual-frequency dielectric resonator antenna with inverse T-shape parasitic strip. In: IEEE Appl. v Comput. Electromagn. Soc. Int. Conf., Apr. 2005, pp. 384–387
20. Chatla, A.B., Bandyopadhyay, S., Maji, V.: A dual band z-shape stepped dielectric resonator antenna for millimeter wave applications. In: Proc. International Conference on Electronics and Communication Engineering (ICECE—2012), Vizag, India, 28–29 April, 2012, pp. 180–184
21. Microwave Studio Version 3. CST GmbH, Darmstadt, Germany

Chapter 8
OCDMA: Study and Future Aspects

Shilpa Jindal and Neena Gupta

Abstract This chapter is on the OCDMA technique that shows light on the present trends of coding and its future aspects. The challenges that are faced by this technology are also documented. This technique employs newly developed three-dimensional codes based on optical orthogonal codes and codes from algebra theory, and their performance is evaluated on two models, Model A and Model B. The results are judged in terms of bit error rate, number of users, eye diagrams, etc. Some experiments performed recently by researchers are also documented.

Keywords OCDMA · Topology · Congruent operators · Optical multiplexer · Optical de multiplexer · Bit error rate · Eye diagram · Quality factor · Data rates

8.1 Introduction

Today's communication networks are changing its structure from asymmetric to symmetric (both upstream and downstream), and their requirement for advanced optical networks has rejuvenated OCDMA research. Hence, to support bandwidth-thirsty applications such as multimedia, high-definition TV, 3D TV, telemedicine, voice, video, and data, there is a need of a transmission media such as optical fiber cable that has capability to support these applications transferring tera bytes of data due to some of the inherent characteristics such as low transmission loss, small size and weight, resistant toward electromagnetic interference and radio frequency interference, high capacity (due to the range of optical carrier frequencies of 10^{13}–10^{16} Hz), and low cost due to large availability of manufacturing material. Due to success in CDMA technology in wireless domain, researchers thought of implementing CDMA in optical domain. By combining the features of two technologies, a good gain over the future networks can be achieved. All optical

S. Jindal (✉) · N. Gupta
PEC University of Technology (Formally Punjab Engineering College), Chandigarh, India
e-mail: ji_shilpa@yahoo.co.in

S. Patnaik et al. (eds.), *Recent Development in Wireless Sensor and Ad-hoc Networks*, Signals and Communication Technology, DOI 10.1007/978-81-322-2129-6_8

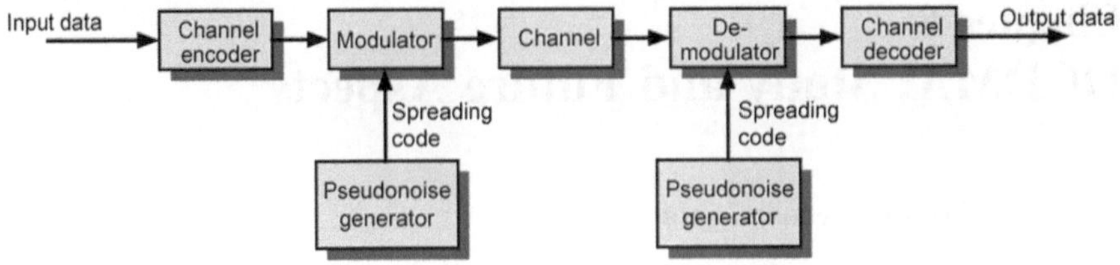

Fig. 8.1 Block diagram of spread spectrum technique [24]

processing achieves high speed as it avoids electrical-to-optical and optical-to-electrical conversion, and CDMA employs inbuilt security and interference avoidance due to spreading bandwidth.

$$\text{OCDMA} = \text{Optical processing} + \text{CDMA}$$

A brief overview of spread spectrum and its types is discussed below.

Spread Spectrum In spread spectrum, the transmission bandwidth is much greater than the baseband message signal bandwidth and the transmission bandwidth is determined by a spreading signal that is independent of the message. Spread spectrum technologies are classified as direct sequence spread spectrum (DSSS) and frequency hopping spread spectrum (FHSS). As shown in Fig. 8.1, the block diagram of spread spectrum technologies consists of one pseudorandom generator that spreads the signal to infinite bandwidth and same generator despreads the signal to original form at receiver section.

8.2 Channel Accessing Technologies

In the literature, many researchers had concentrated on the time division multiple access techniques and wavelength division multiple access techniques. Although these techniques support above high-speed and bandwidth applications, synchronization for time and hardware constraint for wavelength division multiple access techniques hinders its practical applicability. So there is a need to utilize the code division multiple access technique dominantly accepted in today's wireless mobile communication in optical domain. Further choice of particular multiplexing technique evaluates the performance (bandwidth) of any optical network. In multiple access, the number of users shares a common transmission medium to transmit messages. Channel accessing technologies are classified based on the following domains.

- Time division multiple access (TDMA): Here, each channel occupies a preassigned time slot according to some mathematics. To communicate through TDMA, strict synchronization between transmitter and receiver is required.

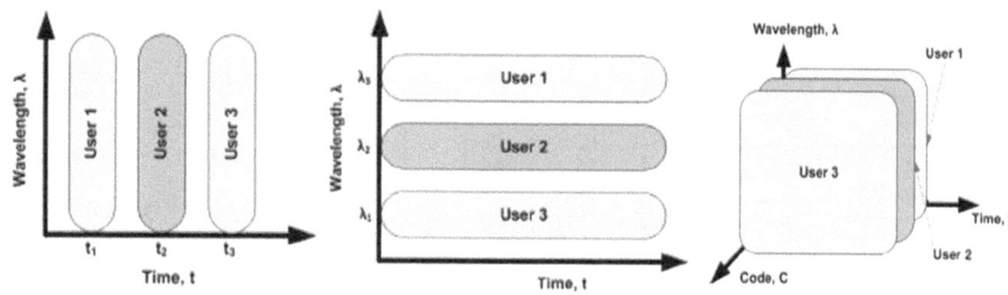

Fig. 8.2 TDMA, WDMA, and CDMA

- Wavelength division multiple access (WDMA): Here, each user is allocated wavelengths and handshaking between users is done through wavelengths. Interfrequency interference is the biggest hurdle in implementing these systems.
- Optical code division multiple access (OCDMA): It is a type of spread spectrum communication in which multiple access is achieved by assigning different, minimally interfering code sequences to different users to communicate in optical domain. Requirement of these code sequences is that these codes should be orthogonal to each other.

Upstream refers to data transmitted from user to server or central location, and downstream refers to data transmitted from server to the user.

In TDMA system, each channel occupies a preassigned time slot as shown in Fig. 8.2.

In WDMA system, each channel occupies a narrow optical bandwidth ≥100 GHz around a center wavelength. WDMA is classified as DWDM and CWDM differentiating based on channel spacing [1, 2].

1. **Dense WDM (DWDM)** can be employed, with reduced channel spacing (ITU-T recommendation G.694.1 defines 43 wavelength channels from 1,530 to 1,565 nm, with a spacing of 100 GHz).
2. Channel spacing for CWDM is slightly loose in comparison with that of DWDM, i.e., 20 nm, which makes CWDM equipment compact and cost effective at the expense of less number of channels (ITU-T recommendation G.694.2).

In OCDMA, network resources are shared among users which are assigned a code instead of time slot like TDMA or a wavelength like WDMA.

OCDMA: Transmission of a signal in a fiber-optic channel is formed by a pseudorandom OCDMA signals encoded using multiple channels. These address sequences should have the following properties. All address sequences should be independent or easily distinguishable of each other's versions, i.e., sequences should have low cross-correlation. All address sequences should be easily identified from its shifted versions, i.e., sequences should have high autocorrelation peak. Figure 8.3 shows the configuration of the OCDMA system. Here, each bit is divided into time periods, wavelengths, and spatial channels in the encoder section. Each user on the OCDMA system has been assigned a unique signature sequence

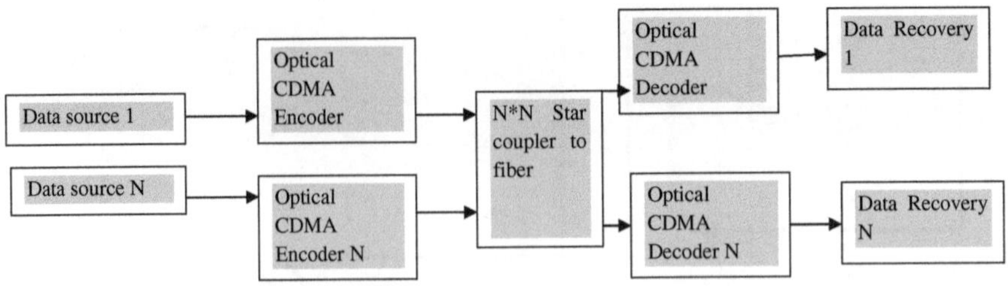

Fig. 8.3 OCDMA block diagram

(address sequences). Data or information to be transmitted is encoded in the encoder based on researcher's code. The encoder of each transmitter transmits one signature sequence for each bit 1 and 0 for binary 0, that is, it is not encoded and is represented using an all-zero sequence. Signature sequence comprising 1 shows the presence of one with corresponding wavelength, time slot, and spatial channel. After encoding, signal is propagated to respective decoder through star coupler whose job is the reverse of that of encoder. Then, the signal is passed to the receiver in the data recovery unit.

8.3 Advantages of OCDMA

The following are some advantages of OCDMA technology over other technologies:

1. Simple, random, and simultaneous access protocol.
2. Bursty network environments.
3. Asynchronous transmission simplifies network management and control.
4. Reduction of optical resources required for timing recovery.
5. Ability to support variable data rates therefore suitable for triple play.
6. No need of strict timing and wavelength control.
7. Efficient utilization of bandwidth.
8. Immunity to various noises.
9. Advantage of combining the large transmission bandwidth of fiber optic.
10. Flexible.
11. Good performance under multiple access conditions.
12. Privacy and security in transmission against interception.

8.4 OCDMA on Timescale

Researchers worked on OCDMA technology, and their significant contribution has been divided by Prucnal on timescale of 5 year each.

In 1980s, the first paper on OCDMA was reported [3, 4]. In 1990s, researchers started shifting their interest from 1D code to 2D codes due to the requirement to increase the user base. 2000s showed the area of performing network experiments in the laboratories. From the literature, the following people have shown major contribution in OCDMA field

1. 1975–1980: Dixon (1976)
2. 1980–1985: Marom, Shedd, Robinson, and Pursley
3. 1985–1990: Froehly, Heritage, and Thurston,
4. 1990–1995: Weiner, Salehi, Leaird, Gelman, Foschini, and Kiasaleh
5. 1995–2000: Yao, Wefers, Brandt-Pearce, Kavehrad, Aazhang, Young, Hinkov, Iversen, Pleiffer, and Elbers
6. 2000–2005: Chang, Sardesai, Gruneet-Jepsen, Purchase, Zheng, Cao, Kavehrad, Nguyen, Dennis, Dutt, Dennis, Tancevski, Fathallah, Kiasaleh, Yegnanarayan, Yu, Kutsuzawa, Chen, Fortier, Kim, and Ben Jaafar
7. 2005–2010: Liang, Cong, Scott, Hernandez, Jian, Kavehrad, Young, Babich, Chan, Inaty, Wei, Patel, Baby, Kwong, Kissing, and Schmuck
8. 2010–till date: Yin, Hongxi, Richardson, David J., and Maninder

8.5 Challenges for OCDMA

Many challenges have to be faced by any technology for its practical implementation and same is true for OCDMA. Few of which are discussed here as follows [5]:

1. Coding Algorithms and Schemes: The properties of codes, i.e., autocorrelation and cross-correlation constants, cardinality, weight, and length of the codes, are few challenging parameters in order to maintain good bit error rate. These are the parameters on which the performance of any OCDMA system is evaluated by judging the multiple access interference (MAI), i.e., by knowing the number of active users that can transmit signal asynchronously without error. In the literature, researchers had proposed various coding algorithms and schemes to generate these pseudorandom signature sequences.
2. Network Architecture: Mapping of particular OCDMA application to a given architecture depends on network scalability, device integration, system cost, and environment robustness. It is the way in which any network is connected. The main topologies are star, tree, mesh, ring, bus, free-space networks, or hybrid which is any combinations of above-said topologies.

Star topology In OCDMA, star topology is normally followed where the number of users increases than the number of input ports and output ports. There are two approaches for star networks. One is that optical combiner and optical splitter along with optical coupler are used to accommodate the increasing number of the subscribers as shown in Fig. 8.4, and other is to employ optical multiplexers and demultiplexers instead of combiners and splitters.

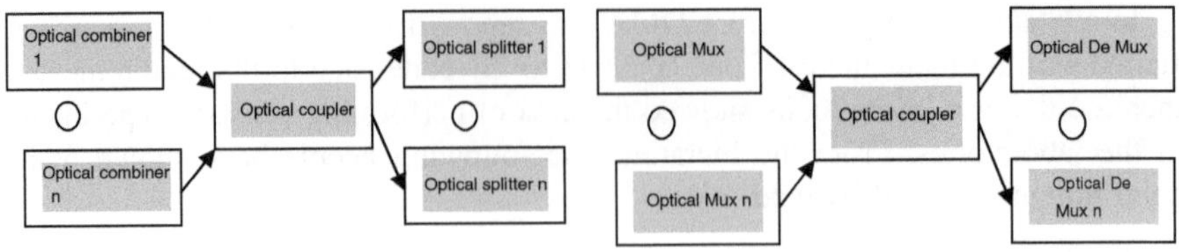

Fig. 8.4 Star topology

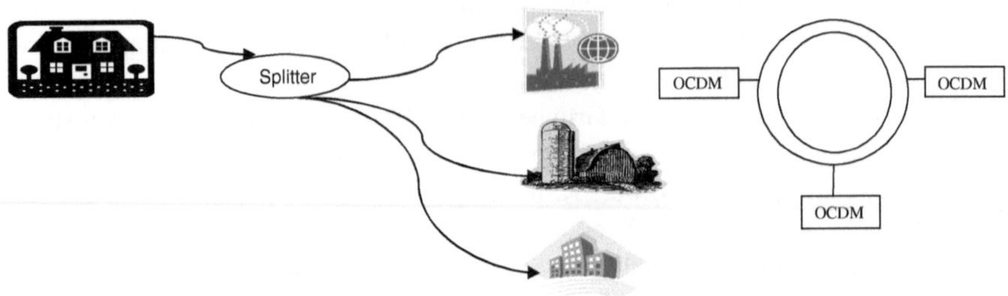

Fig. 8.5 Tree and ring topologies

Ring/Bus Although star topology is mainly used in OCDMA networks, it has the limitation that it can be employed local area networks. To cover wider networks, ring network is suitable. Optical add drop multiplexer is the main component for ring network along with OCDMA node.

Tree For varying traffic demand patterns or due to busty nature of traffic in access networks, multipoint-to-multipoint technology may not be efficient in star topology. For passive optical technology, deploying fiber from central office to fiber to the home, fiber to the curb, and fiber to the office, i.e., for FTTx tree topology, is the most suitable as shown in Fig. 8.5.

Hybrid network This topology consists of implementing combination of above topologies as per the application requirements.

3. Advanced Encoding and Decoding Hardware: The main requirement to implement OCDMA is how to encode and decode the user data that could produce and detect signature sequences efficiently. Hence, encoding in temporal, spectral, spatial, or polarization domain can be done by employing advance optical processing components such as fiber Bragg grating, optical filters, polarization controllers, optical couplers, optical multiplexers, and demultiplexers. Hardware employed for OCDMA directly adds cost. Hence, encoding and decoding architecture is the significant factor in determining the implementation cost of OCDMA system. Encoders and decoders are classified based on coherent and incoherent sources.
4. Simulation and Applications. Simulation is the imitation of the operation of a real-world process or system. Many simulators are available to demonstrate the

OCDMA technology, for example, Rsoft's OPTSIM, Opti wave, MATLAB Optical Toolbox, and DARPA Linux-based simulator. These technology demonstrator programs judge the performance, security, and multiple access interference issues in OCDMA. Main applications of OCDMA lie in the fields of medical, communication, broadcasting, and networking including multimedia, high-definition TV, 3D TV, telemedicine, voice, video, and data.

8.6 OCDMA Codeset Design Issues

Since OCDMA is currently an active area of research, it would be premature to single out just one domain or area of technology. As discussed in the last section, OCDMA has to face many challenges before maturing. That is why a lot of research is being carried out to bring the best design with good performance for future networks in the domains of coding algorithms, network architecture, advance encoding and decoding hardware, simulations, and applications. To inline with the literature, good codeset considering following issues for OCDMA is required for applications in the field of military, medical, and communication.

1. **Length of codes**: If the number of users increases in the 1D and 2D systems, then the code length of OCDMA codes increases many folds, and hence, this increased length of resultant code makes it too long and practically cost ineffective to implement. So third or higher dimensions are considered as a degree of freedom to increase the number of permissible users with allowed BER with practical lengths.
2. **Cardinality**: Cardinality signifies the number of users available or actually the potential users. Assigning codes to various users depends on the coding scheme chosen.
3. **Weight**: The number of pulses high or on defines the weight of the codeset. It should normally be less.
4. **Hardware complexity and cost**: For designing the codes, the existing concept of using polarization as one of the dimension makes system bulky and costly, since there is the requirement of additional components for maintaining polarization of the signal.
5. **Complex Construction Mechanism**: Some codes have construction mechanism that is complicated. To make codeset more popular and practically implementable, there is a need to generate codeset which has simple mathematical modeling for structure.
6. **Performance of the system**: Is dependent on the set of the codes chosen, in most of the available codes; it has been observed that for the same set of code group, different codeset chosen for transmission by same number of users results in variation of system performance in terms of BER, eye diagram.

Hence, coding plays a crucial role in judging the overall performance of OC-DMA system.

8.7 Coding Characteries and Classifications

8.7.1 Coding Characteristics

Signature sequences are characterized based on two constants: autocorrelation constant and cross-correlation constant. Autocorrelation is the cross-correlation of a signal with itself, i.e., the signal can be easily identified from its cyclic shifted versions. Cross-correlation of a signal is defined as that every signal is different or distinguishable from other signals and their shifted versions.

1. Autocorrelation constant (λa)
2. Cross-correlation constant (λc)

In order to define correlation constraints, here we are taking an example of three-dimensional codeset. For two 3D codes X and Y with code weight w, the auto-correlation and cross-correlation have the following properties:

Definition 1 Autocorrelation

For any codeword

$X = \{x_{ijk}\}$ of ($W \times T \times S$, λa, λc) 3D codeset with entries x_{ijk} ε {0.1};

where $0 \leq i \leq W$; $0 \leq j \leq T$; $0 \leq k \leq S$; the periodic autocorrelation λa of X satisfies

$$\sum_{i=0}^{W-1}\sum_{j=0}^{T-1}\sum_{k=0}^{S-1} x_{ijk}\, x_{ij} \oplus t_k = w \, \text{for}\, t = \ 0 \ \text{and} \ \leq \lambda a \, \text{otherwise.} \tag{8.1}$$

where $0 < \gamma < T$ and subscript $\oplus$ denotes modulo T addition.

Definition 2 Cross-correlation

For any two code words X, Y

$X = \{x_{ijk}\}$ & $Y = \{y_{ijk}\}$ of ($W \times T \times S$, λa, λc)

3D codeset with entries x_{ijk} **&** y_{ijk} ε {0.1};

where $0 \leq \text{i} \leq W$; $0 \leq j \leq T$; $0 \leq k \leq S$; the periodic cross-correlation λc of X and Y satisfies

$$\sum_{i=0}^{W-1}\sum_{j=0}^{T-1}\sum_{k=0}^{S-1} x_{ijk}\, y_{\text{ij}} \oplus t_k \leq \lambda c \tag{8.2}$$

where $0 < \gamma < T$ and subscript $\oplus$ denotes modulo T addition.

8.7.2 Coding Classification

Code sequences in OCDMA are classified based on various parameters as shown in Fig. 8.6. Broadly, codes are classified as

1. Classification based on dimensions: (a) 1D, (b) 2D, and (c) 3D
2. Classification based on working principle: (a) coherent and (b) incoherent (unipolar {0,1} codes are used)
3. Classification based on polarity: (a) unipolar and (b) bipolar
4. Classification based on construction: (a) direct and (b) search methods
5. Hybrid OCDMA (any combinations of the above techniques)

(1) Classification based on dimensions: In the literature, codes are named as 1D, 2D, or 3D according to the following dimensions:
(a) One-dimensional: Spreading the sequences in one dimension.
(b) Two-dimensional: Spreading the sequences in two dimensions.
(c) Three-dimensional: Spreading the sequences in three dimensions.

These sequences are spread in either of the following domains: (a) time spreading domain, (b) spectral/frequency hopping domain, (c) spatial domain, and (d) polarization domain as shown in Fig. 8.7 and Table 8.1. Performance of CDMA is based on spreading sequences on certain mathematical models. Some examples of spreading sequences in the literature are zero cross-correlation codes, perfect difference codeset based on Singer algorithm, optical orthogonal codes, quasi-optical orthogonal codes, constant-weight symmetric OOC, constant-weight asymmetric OOC, variable-weight OOC, prime codes, modified prime codes, double-padded modified prime codes, synchronous modified prime codes, time spreading codes, frequency hopping codes, linear congruence codes, quadratic congruence codes, extended quadratic congruence codes, and bipolar codes; some of these codes have good correlation properties, and others have good cardinality [3, 4]. Out of all 1D codes available, the optical orthogonal codes have good performance, but it has limitation to hardware implementation as the number of

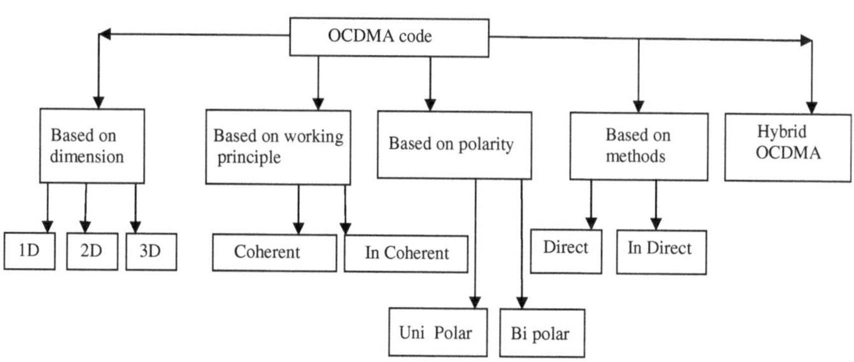

Fig. 8.6 OCDMA code classification

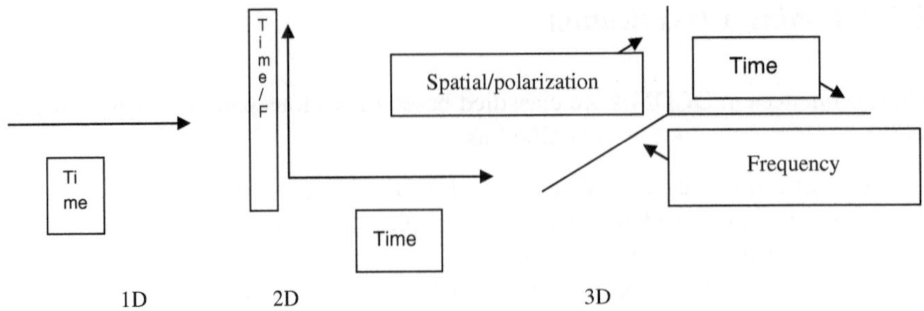

Fig. 8.7 Showing 1D, 2D, and 3D code format [25]

Table 8.1 Dimension of code sequences

1-dimensional		2-dimensional	3-dimensional
Time domain	Wavelength domain	Combination of wavelength/spatial/temporal/polarization domains	
Pulse amplitude	Spectral amplitude	Time and space	Time/wavelength/ polarization
Pulse phase	Spectral phase	W/T (wavelength and time)	Time/wavelength/space

users accessing the network increases. Therefore, there is a need to move to higher-dimensional with increase in the degree of freedom, i.e., 2D and 3D to further improve cardinality and performance of codes for optical networks.

(a) 1D Codes

These are the codes that are spreading the sequences in one domain; it may be temporal or wavelength based on certain mathematics.

Navigating from 1D to 2D

To increase the potential users/spectral efficiency [6], there is a need to migrate the codes from one dimension to higher dimensions as the ratio of code length to code weight grows rapidly as the number of users is increased for a reasonable weight. Hence, for a given pulse width, the data rate decreases or bit rate decreases for 1D codes. To overcome this, spreading is done in both time domain and frequency domain. There are two ways to generate 2D codes either.

(1) 1D codes can be converted to 2D codes by using Chinese remainder theorem or by folding Golomb rulers or
(2) 2D OOCs can be constructed by either time spreading and wavelength hopping patterns by utilizing mathematical modeling to both spreading and hopping algorithms [7]

(b) 2D Codes

These are the codes that are spread in two dimensions. Either of the domains as given in Table 8.1 is considered. These codes are classified as unrestricted codes and restricted codes with the maximum number of users limited by Johnsons bound and singleton bound, respectively.

1. Unrestricted codes: A maximum number of users possess Johnsons bound

$$(\Phi(\Lambda \times T, w, k) \leq\ = [\{(\Lambda/w)\}\{(\Lambda t - 1)/(w - 1)\}\ldots\{((\Lambda \times T) - k)/(w - k)\}] = J_A(\Lambda \times T,\ w,\ k))$$

where Λ is the wavelength, T is the time, w is the weight, and k is the maximum collision parameter.

2. Restricted codes: The codes with some restrictions applied as follows:

Array with one-pulse-per-wavelength (OPPW) restriction: Each row of every (*T) code array in C must have hamming weight = 1

Array with at most one-pulse-per-wavelength (AM-OPPW) restriction: Each row of every (*T) code array in C must have hamming weight ≤1. The following bound is applied for these codes:

$$(\Phi_{\text{AM-OPPW}}(\Lambda \times T, w, k) \leq\ = [\{(\Lambda/w)\}\{T(\Lambda - 1)/(w - 1)\}\ldots\{T(\Lambda\ - k)/(w - k)\}]$$

For special case, $\lambda = w$, and for the case when there is exactly one pulse per wavelength, the bound above reduces

$$\Phi(\Lambda \times T, w, k) \leq\ = T^K$$

which is termed as singleton bound, where Λ is the maximum number of wavelengths, T is the maximum time slot number, w is the weight of the code, and k is the maximum collision parameter. Figure 8.8 shows the example of one restricted AM-OOPW 2D code.

(a) Arrays with one pulse per time slot (OPPTS): Here, each column of every (*T) code array in C is required to have hamming weight = 1.
(b) Arrays with at most one pulse per time slot (AM-OPPTS): Here, each column of every (*T) code array in C is required to have hamming weight ≤1.
(c) *W/T* MPR (multiple pulse per row) FO-CDMA network
(d) 3D Codes

1 dimension and 2 dimensions although generate good cardinality, but still a reasonable number of wavelengths and chip times cannot easily accommodate many users. Therefore, characteristics of light that it can be transmitted through orthogonal polarization as a third dimension can be looked upon to increase the number of potential users in OCDMA network, or the spatial domain can be considered for spreading in the third dimension. Figure 8.9 shows the structure of these code sequences.

	t0	t1	t2	t3	t4	t5
λ1	1	0	0	0	0	0
λ2	0	0	0	0	0	0
λ3	1	0	0	0	0	0
λ4	1	0	0	0	0	0
λ5	1	0	0	0	0	0
λ6	1	0	0	0	0	0

Fig. 8.8 (6 × 6, 5, 1) Restricted AM-OOPW 2D optical orthogonal codes [26]

Fig. 8.9 Example of 1D, 2D, and 3D codeset structure

(2) Classification based on working principle: (a) coherent and (b) incoherent (unipolar {0,1}codes are used)
Coherent encoding is based on the coherent light source usage like mode lock lasers. These sequences are further classified as spectral and temporal phase encoding that manipulates the phase of the signal in spectral or temporal domain respectively. Examples include temporal phase and spectral phase time spreading, whereas incoherent encoding employs intensity modulation/direct detection. Examples are spatial encoding, 2D and 3D wavelength hopping/time spreading/spatial encoding, etc.

(3) Classification based on polarity: Due to success of CDMA in wireless domain, these CDMA codes were tried in optical domain that gave new direction to researchers for future generation networks. Codes employed in wireless domain were bipolar in nature with +1+1−1+1+1+1−1−1 as example where both positive polarity and negative polarity are considered, whereas codes with unipolar nature have a pattern of 0 and 1 like 110100111. To name a few, Fig. 8.10 shows the unipolar code sequences as optical orthogonal codes (OOC), prime codes, and algebraic congruent sequences and bipolar code

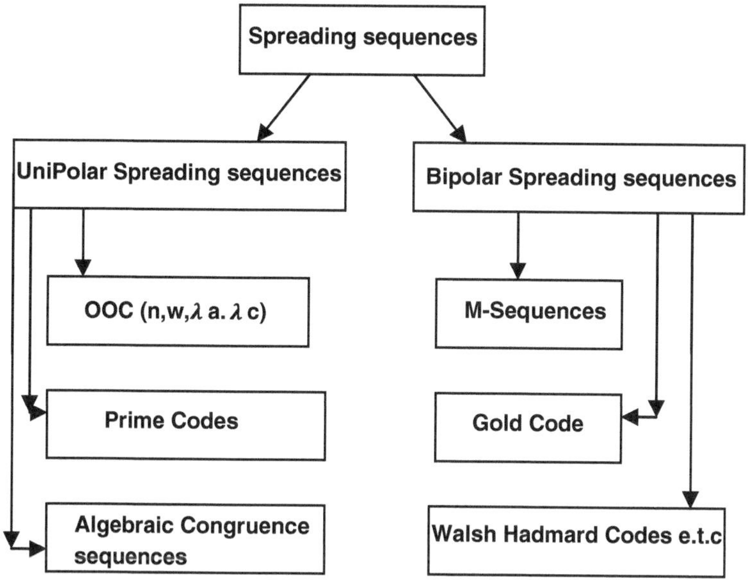

Fig. 8.10 OCDMA spreading sequence tree [11, 13, 14, 25, 27]

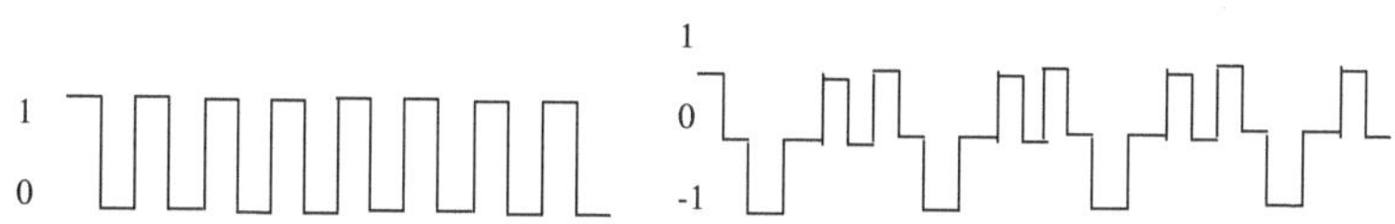

Fig. 8.11 Structure of unipolar and bipolar code sequences

sequences as M sequences, gold code, and Walsh Hadamard codes. Further, Fig. 8.11 shows the structure of unipolar and bipolar code sequences.

(4) Classification based on code construction methods: As given by Salehi, OCDMA code construction methods are classified into two categories: the mathematical modeling (direct method) or generated through computer algorithms (search method) [4]. Detail classification is shown in Table 8.2. The main features of these codes are that they should have good autocorrelation and cross-correlation constraints. In iterative construction method, OOC codes are constructed from existing OOC codes. In greedy algorithm, OOC codes are constructed with general parameters. In projective geometry, finite projections of lines in the geometry are considered as construction grounds. For combinatorial methods, certain algorithms are specified. In block design, there is an arrangement of elements in blocks and every element occurs in blocks and every pair of elements occurs in blocks. Algebraic coding theory involves the operators from algebra like congruent operator with further classifications

Table 8.2 Classification of codes based on dimensions [28]

Direct method	Search method
Projective geometry (recursive construction)	Greedy algorithm
Finite field theory (perfect difference set, finite mobios geometry)	Accelerated greedy algorithm
Design theory	Outer product matrix

as linear congruent, quadratic congruent, hyperbolic congruent, and cubic congruent operator.

(5) Hybrid OCDMA: These OCDMA systems combine any of the above methods and result in hybrid system.

Examples of code sequences are as follows:
A. *OOC (n, w, λa, λc): Optical Orthogonal Codes*

(1) *Constant-Weight Symmetric OOC*: $C = (n, w, \lambda_a, \lambda_c)$ is an optical orthogonal code (OOC) of 0 and 1 sequences of length *"n"* and Hamming weight *"w"* (the number of "1" in each codeword), with periodic autocorrelation of each codeword (λ_a) and the periodic cross-correlation between any two distinct codewords (λ_c). Both λ_a and λ_c must be near ideal in order to distinguish these sequences in the presence of other sequences. For symmetric codes, $\lambda_a = \lambda_c$ For example, Salehi OOC (32, 4, 1, 1) codes are represented as $A =$ [1, 10, 13, 28] and $B =$ [1, 5, 12, 31] with autocorrelation and cross-correlation as 1. These types of codes can be constructed from (a) combinational constructions, (b) algebraic constructions, (c) constructions from finite projective geometry, and (d) recursive constructions.

(2) *Constant-Weight Asymmetric OOC*: When autocorrelation constant is not equal to cross-correlation constant, these types of codes are termed as asymmetric OOC with $\lambda_a \neq \lambda_c$. Although the autocorrelation constraint of OOC acts as a guarantee that each signature sequence is different from its cyclic shift and this property enables the receiver to implement synchronization, there is a requirement of increasing the cardinality of the codes; therefore, relaxation in the constant is considered.

(3) *Variable-Weight OOC*: Here, the third constraint along with previous autocorrelation constant and cross-correlation constant is the weight of the codes. Weight of each codeword is not the same. Further, the quality of transmission service of the user using a codeword with larger weight is higher, and the codeword using a codeword with smaller weight is lower.

Apart from OCDMA, optical orthogonal codes have applications in mobile radio, frequency hopping spread spectrum communication, medicine, radar, and sonar.

B. *Prime Codes*: Prime codes have simple construction mechanism over OOC codes which make them so popular. Examples are basic prime codes, extended

Table 8.3 Modified prime codes for p = 4

S_0 = 1000 1000 1000 1000
S_1 = 1000 0100 0010 0001
S_2 = 1000 0010 0100 0010
S_3 = 1000 0001 0010 0100

prime codes, modified prime codes (asynchronous and synchronous), and double-padded modified prime codes.

(1) *Basic Prime Codes*: It is based on linear congruence operator as shown in the following table. Firstly, choose prime number p over Galois field and construct prime sequence S with s_{mn} as its elements. For example, take p = 5, elements of the GF (5) are denoted as $S_{0,}$ $S_{1,}$ $S_{2,}$ $S_{3,}$ and S_4 and the code sequences generated from linear congruent operator are S_0 = 00000: S_1 = 01234: S_2 = 02413: S_3 = 03142 and S_4 = 04321 that denote the position of high pulse in the sequence of 25 pulses.

(2) Modified Prime Codes: Here, the code sequences from the basic prime codes are expanded in 0 and 1 forms as shown in Table 8.3 for GF (4), i.e., prime umber p = 4.

C. *Algebraic Congruence sequences*: These are the sequences based on algebra theory. Examples are linear congruence, quadratic congruence, extended quadratic congruence hyperbolic congruence, cubic congruence, and costas arrays. Table 8.4 shows the mathematical formulae for various operators that could be implemented in various domains.

Algebraic congruent codes utilize congruent operators within the Galois field to obtain the address sequences of different users for OCDMA system. Jindal et al. proposed 3D codeset with same or different operators, and different sequences could be generated based on Model A and Model B.

Table 8.4 Algebraic congruent operators and their formulas [11, 27]

S. No.	Algebraic congruent operator	Formulae
1.	Linear algebraic congruent operator	$sm(n,a,b) = [m \cdot (n \cdot a + b)][(\bmod p)]$ a = 1 and b = 0
2.	Quadratic algebraic congruent operator	$sm(n,a,b,c) = (m(an^2 \cdot + b \cdot n + c))(\bmod p)$ a = b = 1/2 and c = 0
3.	Hyperbolic algebraic congruent operator	$sm(n) = ((am/n) + b)(\bmod p))$ a, b and c are constants whose value is a = 1, b = 0
4.	Cubic algebraic congruent operator	$sm(n,a,b) = \left(m\left(a+n\right)^3 + b\right)\Big)(\bmod p)$ a = b = 0

8.8 Codeset

In order to generate the 3D codeset, sequences are divided and are placed according to Model A and Model B in Sects. 8.8.1 and 8.8.2, respectively.

8.8.1 Model A Codeset

Figure 8.12 shows the placement of linear congruent operator (LCO), quadratic congruent operator (QCO), hyperbolic congruent operator (HCO), and cubic congruent operator (CCO) based on GF (3), and similarly, figures could be drawn to generate the sequences using GF (5) and GF (7).

8.8.2 Model B Codeset

Figure 8.13 shows the placement of LCO, QCO, HCO, and CCO based on GF (3), and other figures could be inference from the Model B, by placing congruent operators and vice versa, for example, placing QCO for hopping in spectral domain and HCO for encoding in spatial domain according to the requirements.

8.9 OCDMA Components and System Parameters

8.9.1 OCDMA Components

To transfer data in any optical communication systems, using OCDMA technologies, the following components are required to build and evaluate the performance of the system in Optsim software. Light sources; Optical Multiplexer; Encoder that consists of Signal generator, PRBS generator, Electrical signal Generator, External Modulator, Optical Filter, Shift signal, Optical splitter, and Polarization transformer; Optical fiber; Decoder; Optical receiver; Eye diagram analyzer; Signal analyzer; and BER.

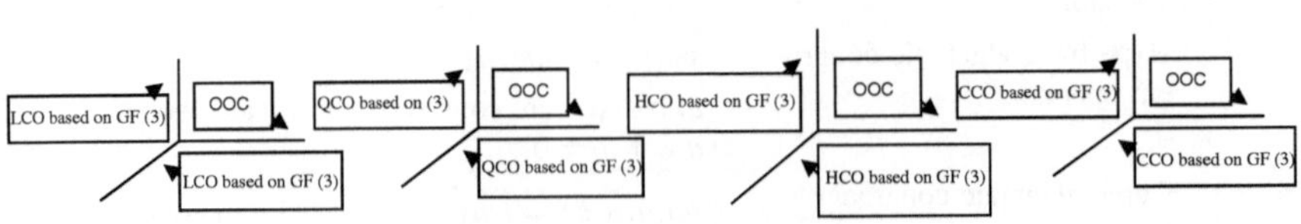

Fig. 8.12 Shows the placement of 3D code based on GF (3) using Model A

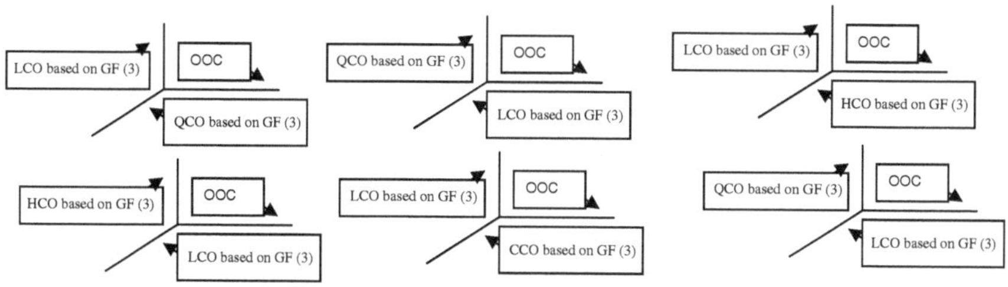

Fig. 8.13 Shows the placement of 3D code based on GF (3) using Model B

(1) **Light sources**: Different types of sources supported are direct modulated laser, mode-locked laser, CW laser, Fabry Perot CW laser, VCSEL, and light-emitting diode (LED).
(2) **Optical multiplexer**: This model represents an optical WDM multiplexer. It accepts multiple optical signals at its input ports and produces a WDM optical signal at its output port which includes all the input WDM optical signals.
(3) **Encoder**: It encodes the signal according to set signature sequence with the help of the following blocks.

Signal generator: It consists of PRBS generator, analog sine generator, electrical signal generator, optical signal generator, custom signal generator, sawtooth generator, frequency sweep generator, and expression signal generator.

PRBS generator: This model generates a binary sequence of several different types. A single model instance may be used to provide multiple pattern outputs, optionally offset from each other, to drive different channels of a WDM or parallel optical bus simulation. Each channel may have its own model instance configured to provide a different pattern than the other model instances.

Electrical signal generator: This model converts an input binary signal into an output electrical signal. The output signal may be specified as either voltage or current. The user parameters are used to configure the electrical signal output. Four different electrical drive types are modeled *On_off, On_off_exp, On_off_ramp, and Raised Cosine*.

Modulator: Different types of modulator supported are as follows: modulator and electroabsorption modulator that modulates an optical signal using electroabsorption modulation.

Optical Filter: This model represents the following types of optical filters: *Fabry Perot*, *Gaussian*, *Raised Cosine*, *Lorentzian*, *Trapezoidal*, *Ideal*, *Custom1*, *Custom2*, and *Custom3*. Each filter type except for the custom types may also be inverted. A wavelength signal whose filtered peak power does not exceed the user-specified drop threshold will not be passed by the filter.

Shift signal: It shifts a signal of any type in time. This model may be used to do one of two primary functions: Delay the output signal relative to the input signal or shift the signal data around in the data structure's array without modifying the signal timing. When signal data are shifted by this block, it wraps around the edges of the data array.

Optical splitter: This model represents an ideal optical splitter. It takes a single input signal and divides it equally among *N* output ports with 1/*N* splitting loss, plus excess loss determined by the transmission model parameter.

Polarization transformer: This model transforms the polarization of the input optical signal(s) according to the specified parameters. This model may be used anywhere in the topology where a specified polarization transformation is desired. The modes of transformation are custom, rotation, linear polarizer, circular left, and circular right.

(4) **Optical fiber**: Different types of fibers supported are Nonlinear Fiber that provides a detailed implementation of propagation of one or more optical channels in a single mode fiber, Bidirectional Nonlinear Fiber (Raman Amplifier) that provides a detailed implementation of a bidirectional fiber with all dispersive, nonlinear, PMD, and Raman Effect.
(5) **Decoder**: It consists of all the components as present in the encoder but with inverse signature sequence.
(6) **Optical receiver**: It consists of compound optical receiver that detects an optical signal and produces an amplified electrical signal and photodetector model either the PIN or APD photodetector.
(7) **Eye diagram analyzer**: It plots eye diagram of a signal and is used to display the eye diagram of a signal at the node connected to its input port(s).
(8) **Signal analyzer**: It displays signal waveforms of a signal at the node connected to its input port(s). It has various settings which allow the user to customize the signal waveform display.
(9) **Bit Error Tester**: It computes BER, Q factor, and eye properties such as the height, width, area, and extinction ratio of an electrical signal. The BER may be calculated using either a quasi-analytical or Monte Carlo algorithm depending on the nature of the dominant noise sources in the simulation. The bit error rate (BER) of a fiber link is the most important measure of the faithfulness of the link in transporting the binary data from transmitter to receiver.

8.9.2 System Parameters

Figure 8.14 [11, 12] shows the structure of OCDMA system based on 3D space/wavelength/time codes supporting algebraic congruent operator. Here, each bit is divided into time periods, wavelengths, and spatial channels. Each user on the OCDMA system has been assigned a unique signature sequence. The encoder of each transmitter transmits one signature sequence for each bit 1 and 0 for binary 0, that is, it is not encoded and is represented using an all-zero sequence. Signature sequence comprising 1 shows the presence of one with corresponding wavelength, time slot, and spatial channel. Encoder 1 through U spreads the user sequences by

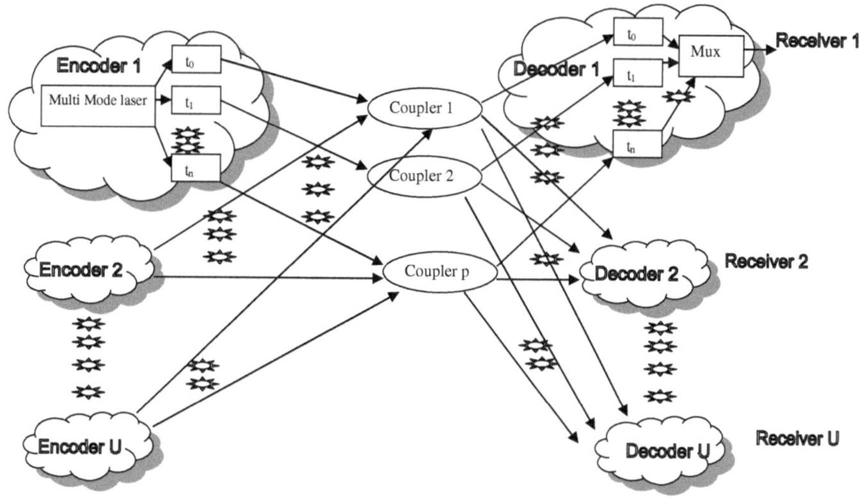

Fig. 8.14 System description for 3D space/wavelength/time codes using algebraic congruent operator (Models A and B)

utilizing wavelength and time slot combinations as depicted by the codeword sequences. Further, these codes are transferred to different couplers using space domain, and the corresponding decoder receives the signals and interprets it. Using this technique, bandwidth of the resultant system is greater than the unspreaded sequence.

While designing any OCDMA system, the parameters such as bit rate, bit period, chip period, laser wavelength, rep rate of source, peak power of laser, delta or channel spacing, number of lasers, combiner/Mux, combiner loss, pattern type, pattern length, bandwidth, fiber attenuator, code length, code weight, polarization, PRBS, measurements, and fiber length should be calculated as shown in Table 8.5. Calculation of chip time with varying bit rate for 1, 2, 2.5, and 5 Gbps system is done as follows: Chip time

(1) For 1 Gbps system,	(2) For 2 Gbps system,
Bit rate: 1e9; time slot = 6 bits	Bit rate: 2e9; time slot = 6 bits
Bit period: 1/bit rate = 1e−9	Bit period: 1/bit rate = 5e−10
Time chip = bit period/time slot = 1e−9/6	Time chip = bit period/time slot = 5e−10/6
So, chip time = 1.6 e−10	So, chip time = 8.3 e−11
(3) For 2.5 Gbps system,	(4) For 5 Gbps system,
Bit rate: 2.5 e9; time slot = 6 bits	Bit rate: 5e9; time slot = 6 bits
Bit period: 1/bit rate = 4e−10	Bit period: 1/bit rate = 2e−10
Time chip = bit period/time slot = 4e−10/6	Time chip = bit period/time slot = 2e−10/6
So, chip time = 6.6 e−11	So, chip time = 3.3 e−11

Table 8.5 System parameters

S. No.	Parameter	Value (example)
1.	Bit rate	1e9
2.	Bit period	1.0e−9
3.	Chip period	1.0e−9/3
4.	Laser wavelength	λ_1 = 1,550.0e−9 m
		λ_2 = 1,550.8e−9 m
		λ_3 = 1,551.6e−9 m
5.	Rep rate of source	1e9
6.	Peak power of laser	0.003w
7.	Delta or channel spacing	0.8e−9 (DWDM)
8.	No. of lasers	3
9.	Combiner/mux	3 × 1
10.	Combiner loss	3 dB
11.	Pattern type	PRBS
12.	Pattern length	28 bits
13.	Bandwidth	1.0e−10 m
14.	Fiber attenuator	−15 dB/Km
15.	Code length	According to code
16.	Code weight	According to code
17.	Polarization	According to code
18.	PRBS	According to code
19.	Measurements	Eye diagram, Q factor and bit error rate
20.	Fiber length	According to system

8.10 3D OCDMA Simulation Results

Table 8.5 shows the practical parameters that were taken while simulating the proposed 3D codeset based on Models A and B using cubic and linear congruent operators with GF (5). Proposed system has 5 operating wavelengths in C band, i.e., λ_1 = 1,550.0e−9 m, λ_2 = 1,550.8e−9 m, λ_3 = 1,551.6e−9 m, λ_4 = 1,552.4e−9 m, and λ_5 = 1,553.2e−9 m with repetition rate = variable and peak power = 1.0e−3 w of MLL (laser). And delta = 0.8e−9 (i.e., spacing between the wavelength) is based on dense wavelength division multiplexing. Figures 8.15 and 8.16 show the snapshots of 3D OCDMA in OPTSIM simulation software for Model A and Model B. This schematic evaluates the 3D OCDMA link with encoding/decoding based on Model A (cubic congruent operator for spectral and spatial domains) supporting 4 users and Model B (linear congruent operator for spatial domain and cubic congruent operator for spectral domain) supporting five users within Galois field GF (5). Both these models use same OOC for spreading in temporal domain.

In this section, in-depth analysis is done for both the models, i.e., Model A and Model B with linear congruent operator, cubic congruent operator, hyperbolic

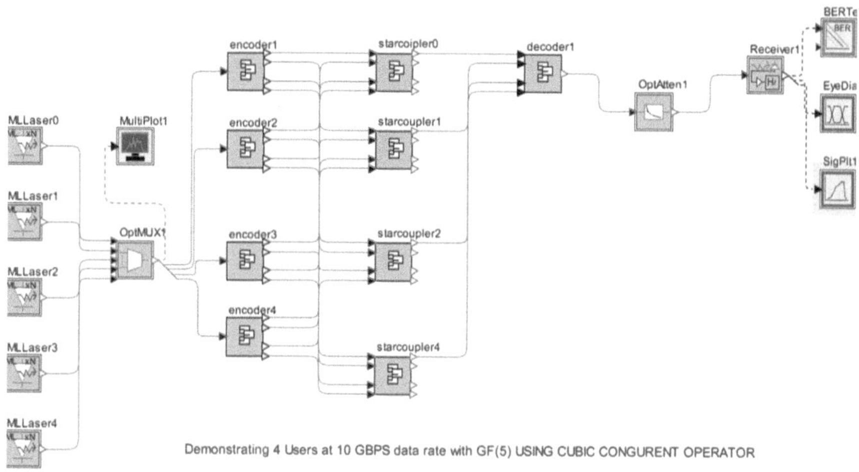

Fig. 8.15 Technology demonstrator of 3D OCDMA system Model A

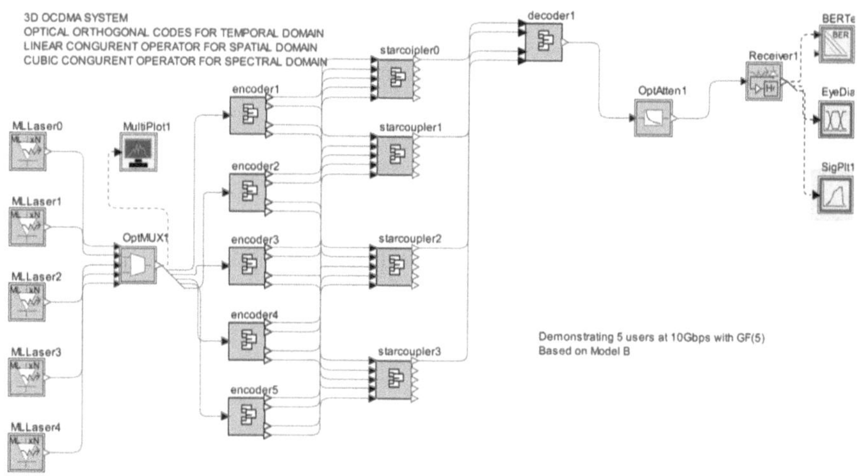

Fig. 8.16 Technology demonstrator of 3D OCDMA system based on Model B [10]

congruent operator, and quadratic congruent operator with varied number of active users. The comparison is drawn between these models based on cardinality, BER versus varying attenuation, bit error rate versus received power at variable data rates and eye diagram and signal strength.

1. **Cardinality**: Cardinality is defined as the number of users supported by the OCDMA system. Based on the cubic and linear congruent operators for Model A,

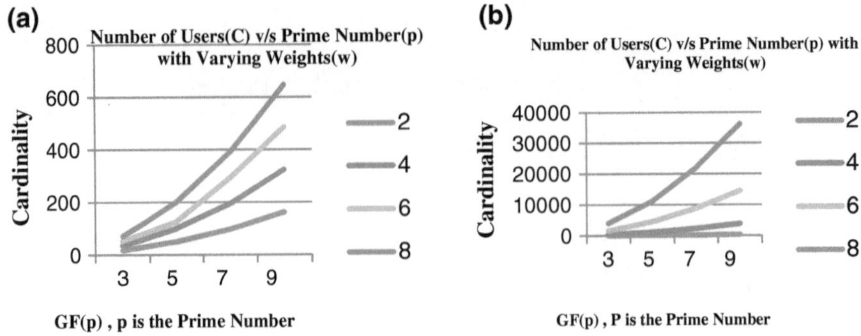

Fig. 8.17 Number of users (*C*) versus prime number (*p*) with varying weights (*w*) for Model A and Model B

$$C = p^2 \times w \tag{8.3}$$

Here, in Eq. 8.3, c is the cardinality, p is the prime number as given by Galois field GF (p) for cubic and linear algebraic congruent operators, and w is the weight of the temporal domain codes. In this simulation work, $p = 5$ and w is 4, so the cardinality in this case is 100. Figure 8.17 shows the cardinality c versus p with varying weight.

For Model B, the cardinality is defined by the following equation:

$$C = p^2 \times w^2 \times (w - 1) \tag{8.4}$$

Here, in Eq. 8.4, c is the cardinality, p is the prime number as given by Galois field GF (p), and w is the weight of the temporal domain codes. In this simulation work, $p = 5$ and w is 4, so the cardinality in this case is 1,200.

2. **BER versus varying Attenuation.**
   Figures 8.18 and 8.19 show BER versus attenuation graphs for Model A and Model B. Here, we have compared the transmission of four and five users based on two models, i.e., Models A and B with varying bit rates and attenuation at the front end of the receiver for 3D OCDMA system. It is observed that for both the models, the BER rate increases with increase in either the data rate or the attenuation. Hence, conclusion can be drawn that for prescribed number of active users, what will be permissible data rate with calculated values of attenuation that can be employed.
   Based on Galois field (5), we were able to simulate proposed codeset based on optical orthogonal codes and codes from algebra theory in OPTSIM simulation software. Results are judged in terms of BER versus attenuation with varying data rate. The simulation results of the OCDMA transmission system validate the feasibility of research.

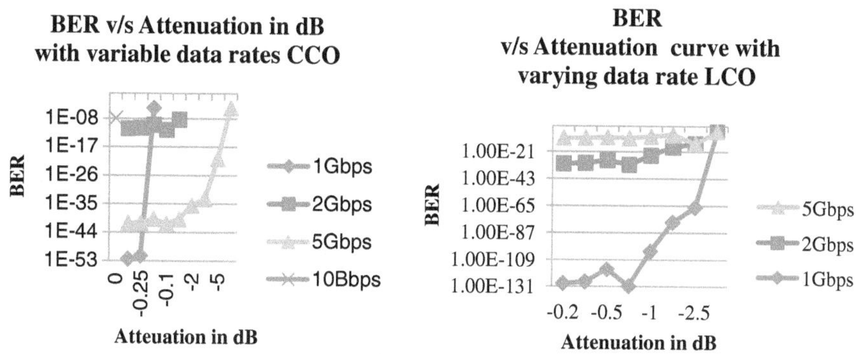

Fig. 8.18 BER versus attenuation curve with varying data rate for cubic congruent operator and linear congruent operator for Model A

Fig. 8.19 BER versus attenuation curve with varying data rate for 5 users based on Model B [25]

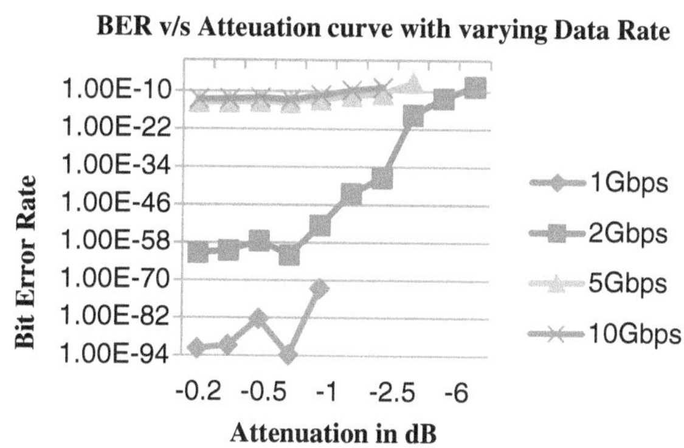

3. Performance comparison in terms of bit error rate versus received power for Model A and Model B at variable data rates

In this section, the comparison of both the models, i.e., Model A and Model B, is shown in terms of bit error rate versus received power at −26, −28, −30, and −32 dBm using algebraic congruent operators as linear congruent operator, cubic congruent operator, quadratic congruent operator, and hyperbolic congruent operator. Accordingly, models are termed as Model A (LC), Model A (CC), Model B (QC-HC), and Model B (LC-CC). Hence, for comparison, two types of each model are simulated, and then, the results are drawn at variable data rate, i.e., at 1 Gbps and 10 Gbps.

(a) At 1 Gbps

The analysis is carried out at 1 Gbps data rate for all the models. The received power is considered as low as −32 dBm with variable number of active users. The measured values are shown in Table 8.6 and are plotted in Fig. 8.20 as graph.

Table 8.6 Bit error rate versus received power at different models at 1 Gbps data rate

Model power (dBm)	Model A (LC)	Model A (CC)	Model B (QC-HC)	Model B (LC-CC)
−26	1.25e−304	5.90e−222	8.63e−279	6.44e−266
−28	1.30e−250	4.29e−99	5.08e−220	4.41e−210
−30	4.80e−192	4.29e−99	6.04e−151	5.95e−151
−32	2.22e−131	1.83e−55	1.16e−91	2.30e−94

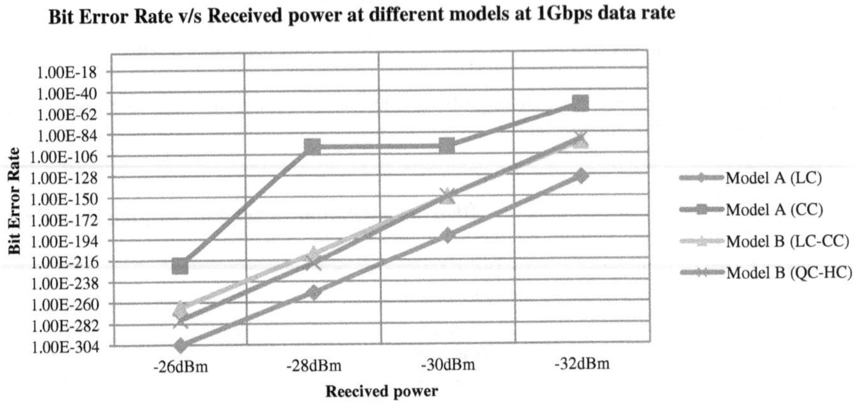

Fig. 8.20 Bit error rate versus received power at different models at 1 Gbps data rate

The results depict that Model A (CC) and Model B (LC-CC) support transmission of maximum four and five number of receivers, respectively, till −32 dBm received power, and if the power increases beyond this value, the BER falls drastically to unacceptable level because of losses and high multiple access interference. Model A (LC) and Model B (QC-HC) support transmission till -32 dBm with maximum thirteen and eighteen number of active users, respectively, with BER greater than 10e−9 at 1 Gbps data rate.

From the values of BER in Table 8.6, it is observed that the code corresponding to receiver 1 shows all values in acceptable range even though the received power is decreased from −26 dBm to −32 dBm at 1 Gbps data rate. So it is recommended to use Model A (LC) and Model B (QC-HC) at −32 dBm received power with acceptable BER for the applications that employ higher number of users over the system that requires few users where Model A (CC) and Model B (LC-CC) can be preferred. Hence, one of each model supports both small and large number of users.

(b) At 10 Gbps

The models are analyzed at very high data rate, i.e., 10 Gbps, the performance in terms of bit error rate further decreases when the data transmitted per user are increased to 10 Gbps. This system has an acceptable BER till −28 dBm received power for Model A (LC) and Model B (QC-HC). The operation of other two

Table 8.7 Bit error rate versus received power at different models at 10 Gbps data rate

Model power (dBm)	Model A (LC)	Model B (QC-HC)
−26 dBm	3.07e−11	3.30e−12
−28 dBm	6.02e−08	3.42e−09

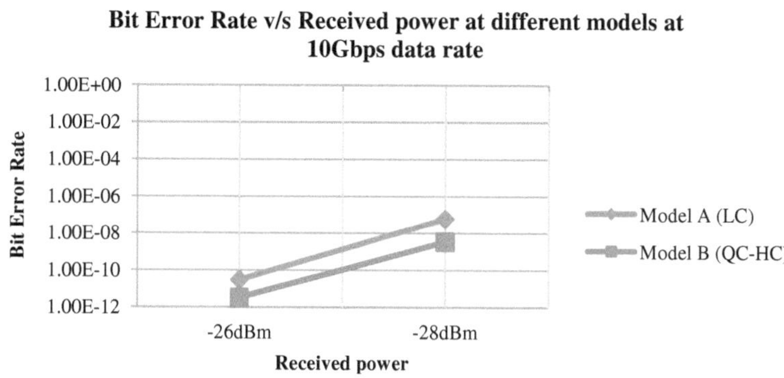

Fig. 8.21 Bit error rate versus received power at different models at 10 Gbps data rate

models fails at 10 Gbps. The BER values are tabulated in Table 8.7 for further reference, and graph is shown in Fig. 8.21.

(4) **Performance comparison in terms of eye diagram and signal strength**

Further, the analysis of these models is done using the spectrum analyzer, and the output is captured in terms of eye diagram and signal strength at variable data rate. The graphs shown here are at −26 dBm received power, and other outputs at other received powers are omitted to avoid repeating. The analysis is carried out in four sections as follows:

(a) Model A (lC),
(b) Model A (CC),
(c) Model B (Quadratic–Hyperbolic Congruent),and
(d) Model B (Quadratic–Hyperbolic Congruent).

The analysis is carried out at 1 Gbps and 10 Gbps data rates.

(a) **Model A (lC)**

The analysis is carried out at 1 Gbps with −26 dBm received power. Figures 8.22 and 8.23 show the eye diagrams and signal strength for input signal and the output signal at 1 Gbps, and Figs. 8.24 and 8.25 show the eye diagrams and signal strength for input signal and the output signal at 10 Gbps, respectively.

It has been observed from the above figures that as the data rate increases, there is a degradation in the output values both in terms of eye diagram and in terms of signal strength diagrams at −26 dBm received power.

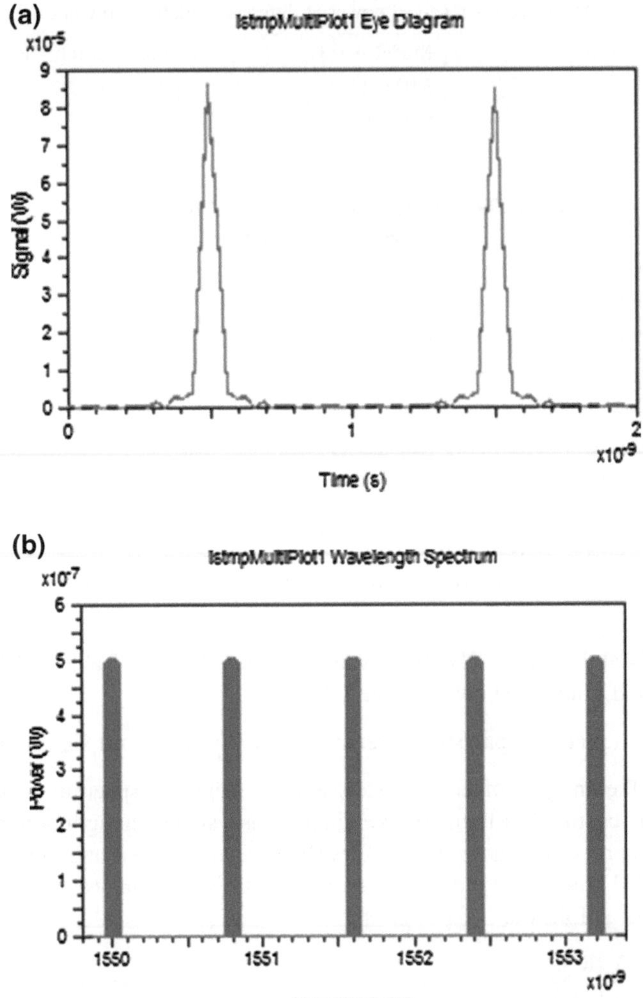

Fig. 8.22 **a** Eye diagram for input. **b** Signal strength for input signal at 1 Gbps

(b) **Model A (CC)**

In this section, Model A (CC) is analyzed at 1 Gbps and 10 Gbps data rates. Figures 8.26 and 8.27 show the eye diagrams and signal strength for input signal and the output signal at 1 Gbps, and Figs. 8.28 and 8.29 show the eye diagrams and signal strength for input signal and the output signal at 5 Gbps, respectively.

It has been observed for this model also that as the data rate increases, there is degradation in the output values both in terms of eye diagram and in terms of signal strength diagrams at −26 dBm received power. The results are shown till 5 Gbps data rate.

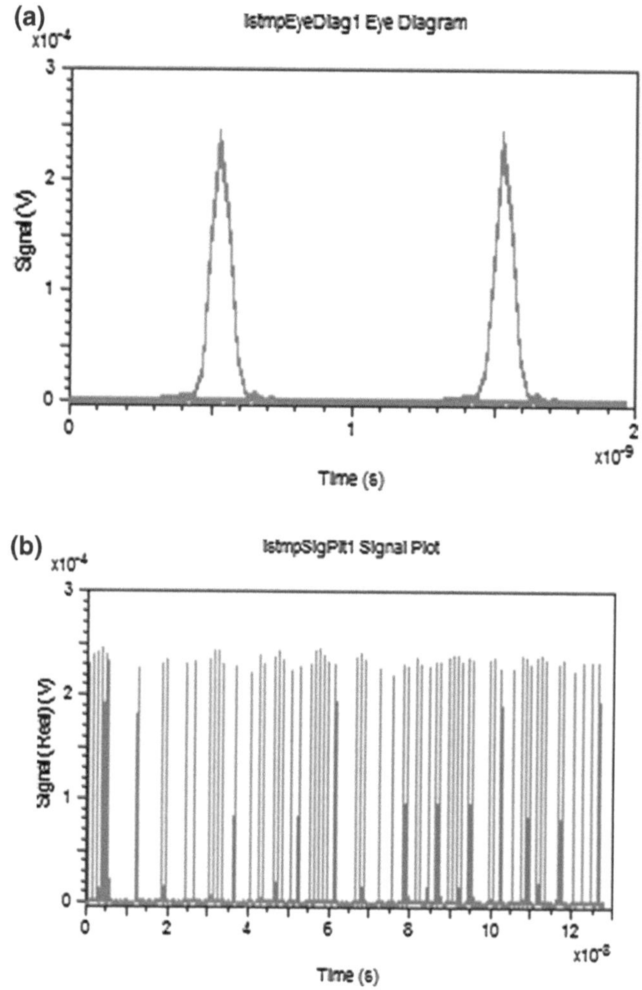

Fig. 8.23 **a** Eye diagram at −26 dBm. **b** Signal strength at −26 dBm at 1 Gbps

(c) Model B (Quadratic–Hyperbolic Congruent)

In this section, Model B (quadratic–hyperbolic congruent) is analyzed till 10 Gbps data rate and Figs. 8.30 and 8.31 show the input values and Figs. 8.32 and 8.33 show the output at 10 Gbps data rate.

(d) Model B (Cubic–Linear Congruent)

Similarly, Model B (cubic–linear congruent) is analyzed for eye diagram and signal strength diagrams till 5 Gbps data rate, and the diagrams at the input and output are shown for 1 Gbps and 5 Gbps data rates at −26 dBm received power at Figs. 8.34, 8.35, 8.36, and 8.37, respectively.

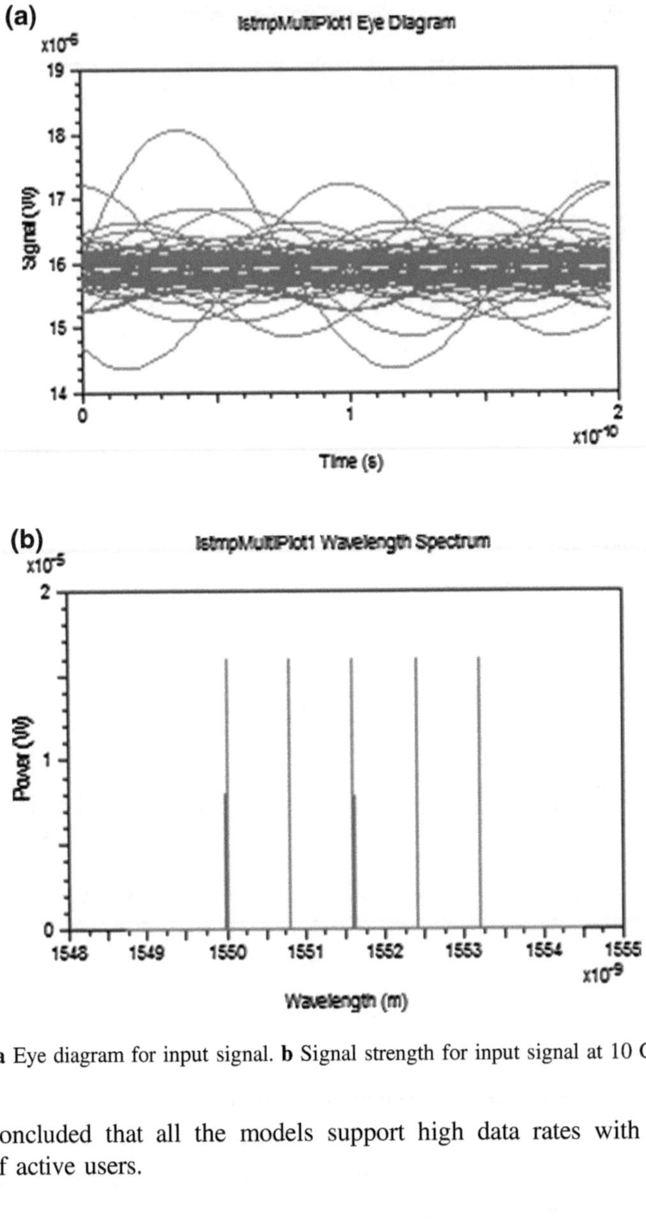

Fig. 8.24 **a** Eye diagram for input signal. **b** Signal strength for input signal at 10 Gbps

It is concluded that all the models support high data rates with prescribed number of active users.

8.11 Technology Demonstrator/Software Support

These are the software programs that help to build and evaluate the performance of optical systems. Example includes OPTSIM, OPTISYSTEM, SIMULINK, and Optical Toolbox in MATLAB [8, 13, 14].

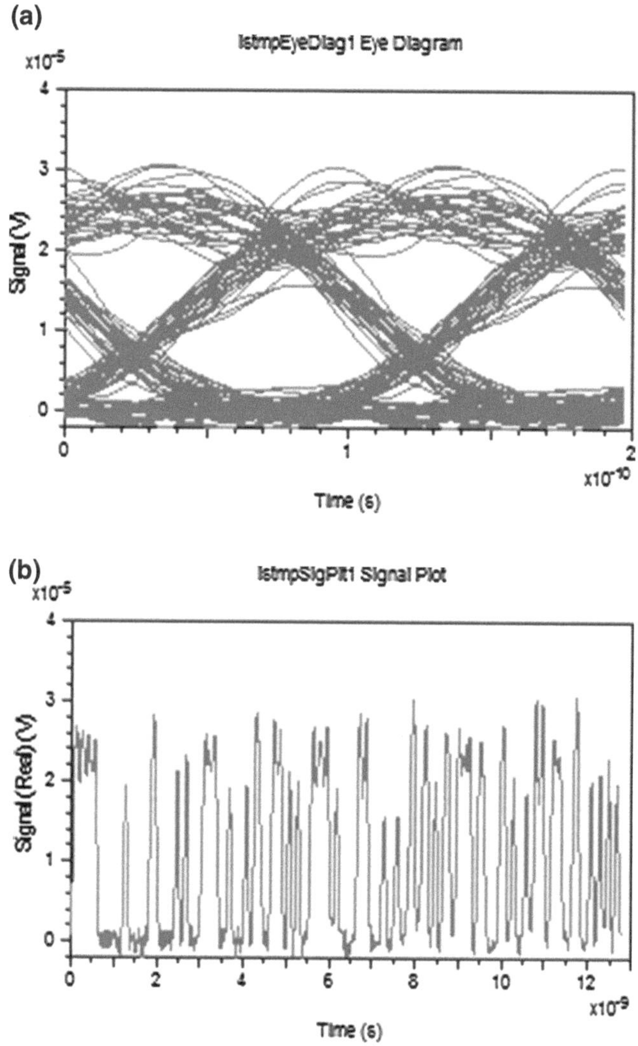

Fig. 8.25 **a** Eye diagram at −26 dBm. **b** Signal strength at −26 dBm at 10 Gbps

1. OPTSIM is an advanced optical communication system simulation package designed for professional engineering and cutting-edge research of WDM, DWDM, TDM, CATV, optical LAN, and parallel optical bus, and other emerging optical systems in Telecom and Data com, and other applications. It is designed by Rsoft design group. The twin simulation engines support two complementary simulation approaches:

(a) The block-mode simulation engine

(b) The sample-mode simulation engine

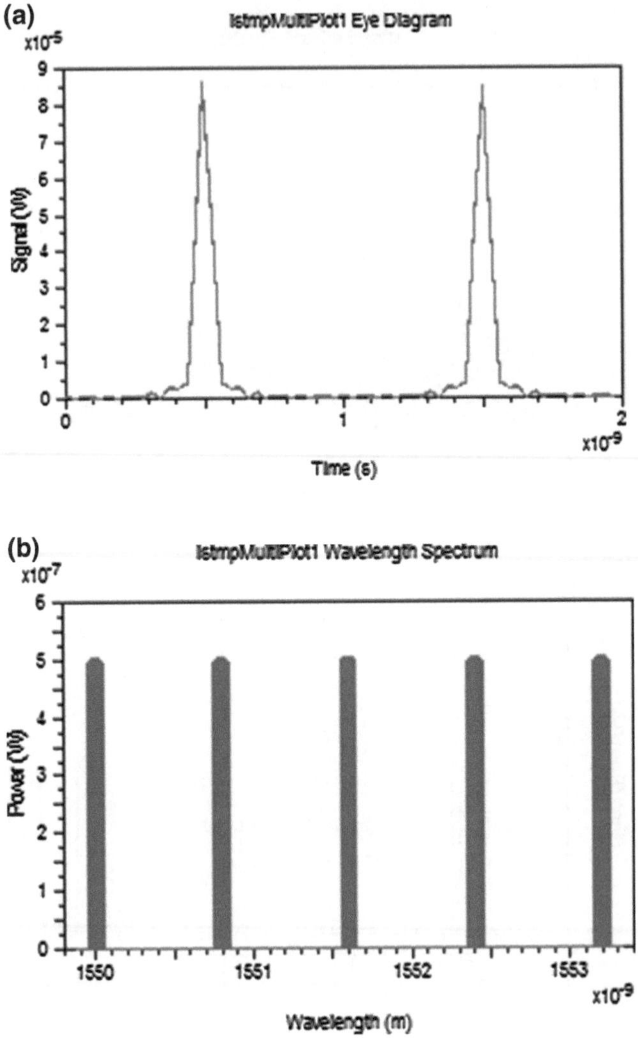

Fig. 8.26 **a** Eye diagram for input. **b** Signal strength for input at 1 Gbps

(a) The block-mode simulation engine: It performs simulations in which the signal data passed between components represent the entire simulated time in one block of data.
(b) The sample-mode simulation engine: It performs simulations in which the signal data passed between components represent a single sample or time step at a time.

Through the flexible user model capabilities provided with OPTSIM, users can create models that operate on signal data as a sample at a time or in large blocks, in either the time domain or frequency domain, for both the modes.

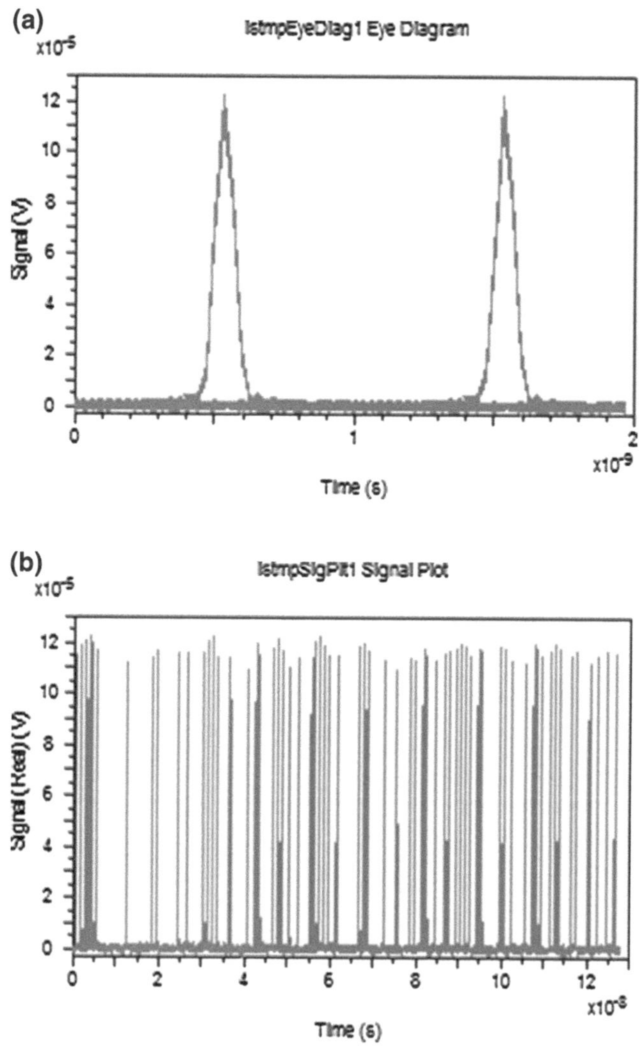

Fig. 8.27 **a** Eye diagram at −26 dBm. **b** Signal strength at −26 dBm at 1 Gbps

2. OPTISYSTEM is an innovative, rapidly evolving, and powerful software design tool that enables users to plan, test, and simulate almost every type of optical link in the transmission layer of a broad spectrum of optical networks from LAN, SAN, and MAN to ultra-long haul. It offers transmission layer optical communication system design and planning from component to system level and visually presents analysis and scenarios.
3. SIMULINK and Optical Toolbox in MATLAB: Optical Fiber Toolbox (OFT) provides functions for fast automatic calculation of guided modes in simple optical fibers.

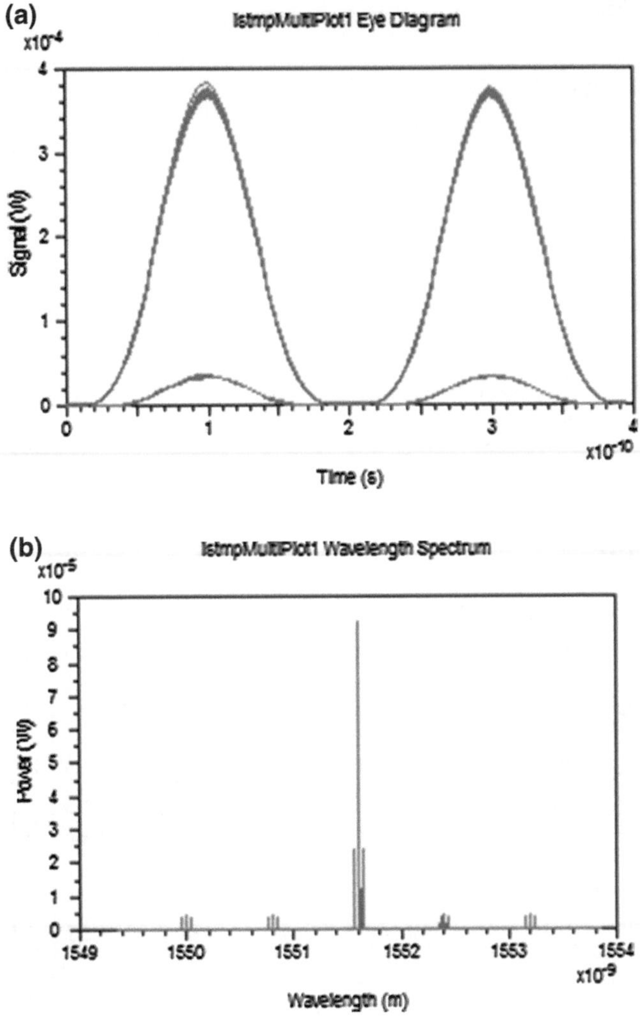

Fig. 8.28 **a** Eye diagram for input. **b** Signal strength for input at 5 Gbps

8.12 Experimental Analysis

1. In 2001, Université Laval, Québe, Canada, demonstrated 16 user experiments with 1.25 Gb/s per user with FFH-OCDMA signal transporting 80 km by using fiber Bragg grating array encoding [15].
2. In 2006, OCDMA Network Demonstration Experiment with star structure was performed by Princeton University in America. It was operated at 115G chips/s by implementing 2D wavelength time-incoherent encoding with single user bit rate of 6 Gb/s and 10^{-13} BER [16].

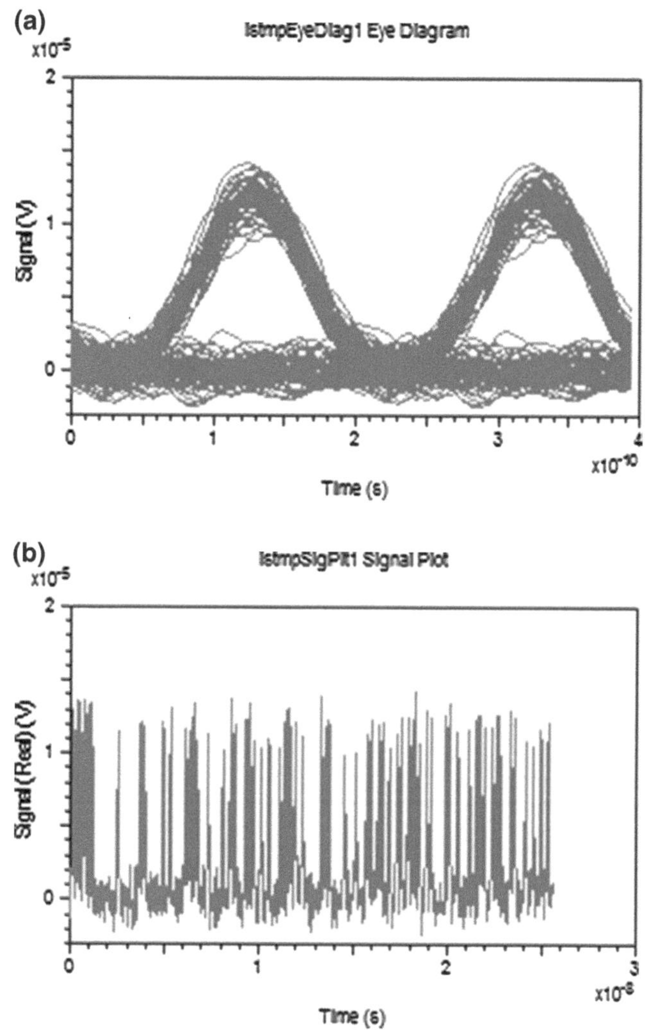

Fig. 8.29 **a** Eye diagram at −26 dBm. **b** Signal strength at −26 dBm at 5 Gbps

3. In 2006, OCDMA LAN protocol 2D wavelength–time encoding with 6 users at 625 Mb/s data rate per user was reported by University of Southern California in USA [9].
4. In 2006, Japan National Institute of Information and Communication Technology demonstrated a cost-effective field trial of 111-km transmission of 3 WDM*10-OCDMA * 10.71 GB/s sharing a single multiport encoder in central office and tunable decoders in ONU. A truly asynchronous WDM/OCDMA access network was implemented with 10 users per wavelength with BER $<10^{-9}$ without forward error correction and optical thresholding [17].

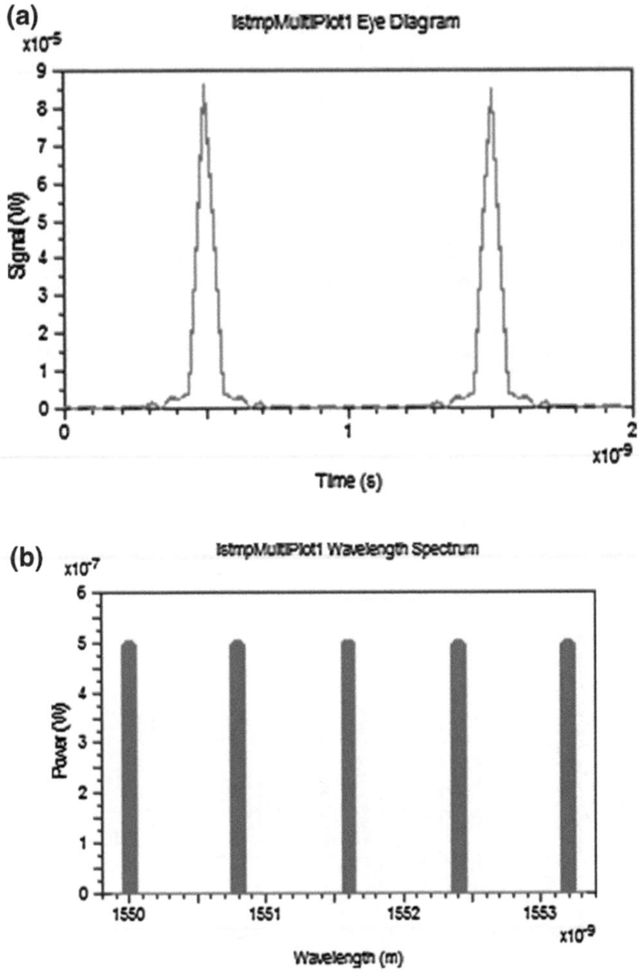

Fig. 8.30 **a** Eye diagram for input signal. **b** Signal strength for input signal at 1 Gbps

5. In 2005, University of California (Davis) in America demonstrated error-free transmission with SPECTS set up with 6 simultaneous users at 10 Gb/s/user data rate with Walsh code length up to 64 chips [18].
6. In 2005, University of California (Davis) in America demonstrated an eight-user 9 GB/s/user time-slotted SPECTS (spectral phase-encoded time spreading) OCDMA test bed. The test bed utilizes a single two-dimensional (2D) phase modulator for encoding multiple channels, each with a unique 64-chip Walsh code. The test bed operated error free with up to six users and at a bit error rate BER $<10^{-9}$ for eight simultaneous users [19].

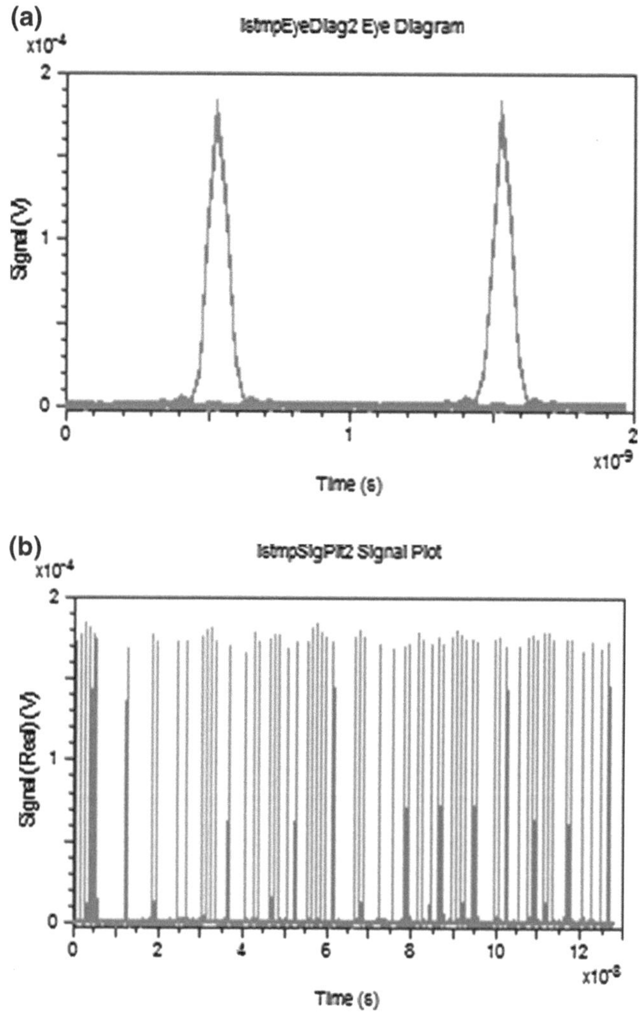

Fig. 8.31 **a** Eye diagram at −26 dBm. **b** Signal strength at −26 dBm at 1 Gbps

7. In 2006, University of California (Davis) in America set up SPECTS (spectral phase-encoded time spreading) LAN experiment with 32 simultaneous users with each user data rate of 10 Gb/s. Tests resulted in BER 10^{-11} with FEC and 10^{-9} without FEC up to 28 users [20].
8. In 2005, DARPA-funded project demonstrated four-user OCDMA system operating at 2.5 Gb/s BER 10^{-11} [21].
9. In 2009, PLA University of Science and Technology, China, demonstrated two-user 2.5 Gbps 100 km OCDMA Transmission Experiment using EPS-SSFBG En/decoder with BER $<10^{-12}$. Further, it was concluded that by using better

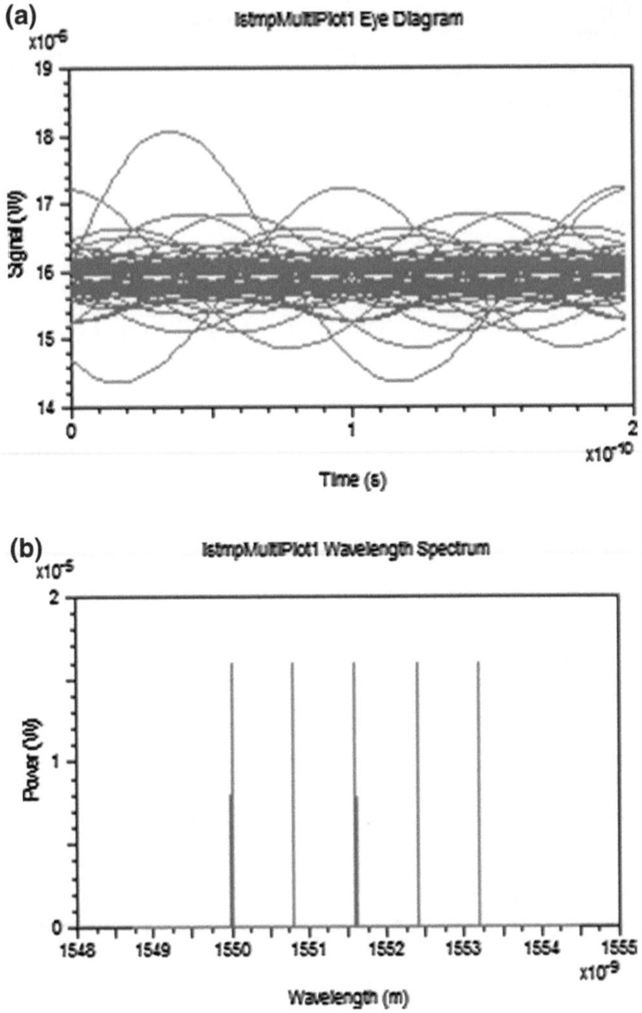

Fig. 8.32 **a** Eye diagram for input signal. **b** Signal strength for input signal at 10 Gbps

optical source and optical thresholding, device system could accommodate more users with larger distances [22].

10. In 2005, DARPA-funded project encoded 1 Gbit/s, 11-Gchip/s data with an 11-chip, 4-wavelength, weight-4 (per polarization) code using free-space, fiber delay lines, and polarization beam combiners. This type of coding increased the number of potential users, by a factor of approximately 2 [23].

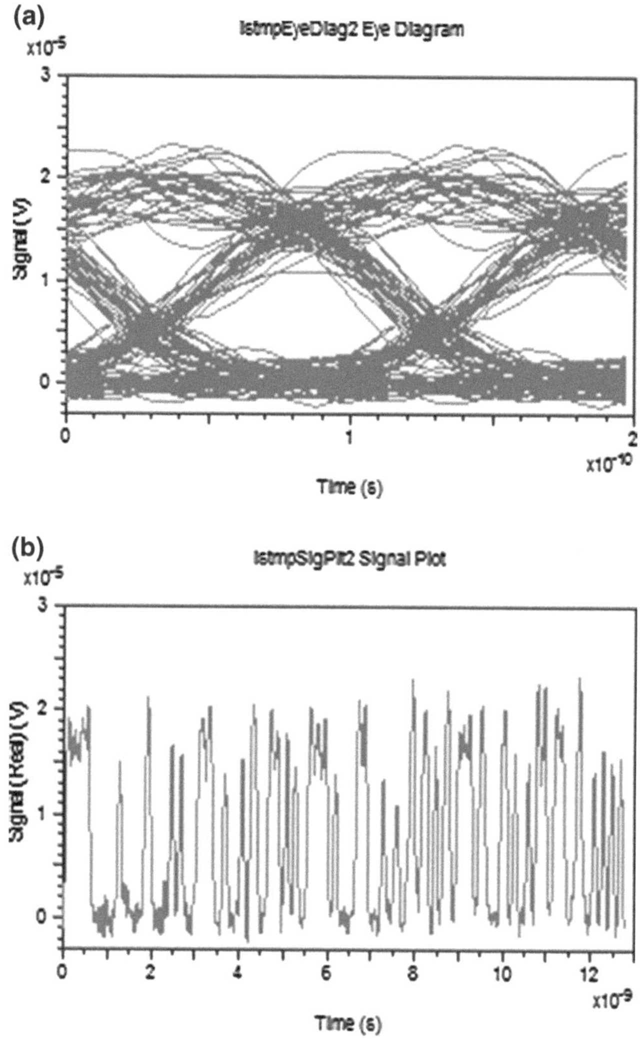

Fig. 8.33 **a** Eye diagram at −26 dBm. **b** Signal strength at −26 dBm at 10 Gbps

8.13 OCDMA Applications

The applications of the OCDMA technology are broadly classified in terms of catering triple play applications, i.e., whether voice or video or data are the priority for establishing a channel. Accordingly, applications are classified under these two heads:

1. Application of OCDMA in communication systems.
2. Applications of OCDMA in data networks.

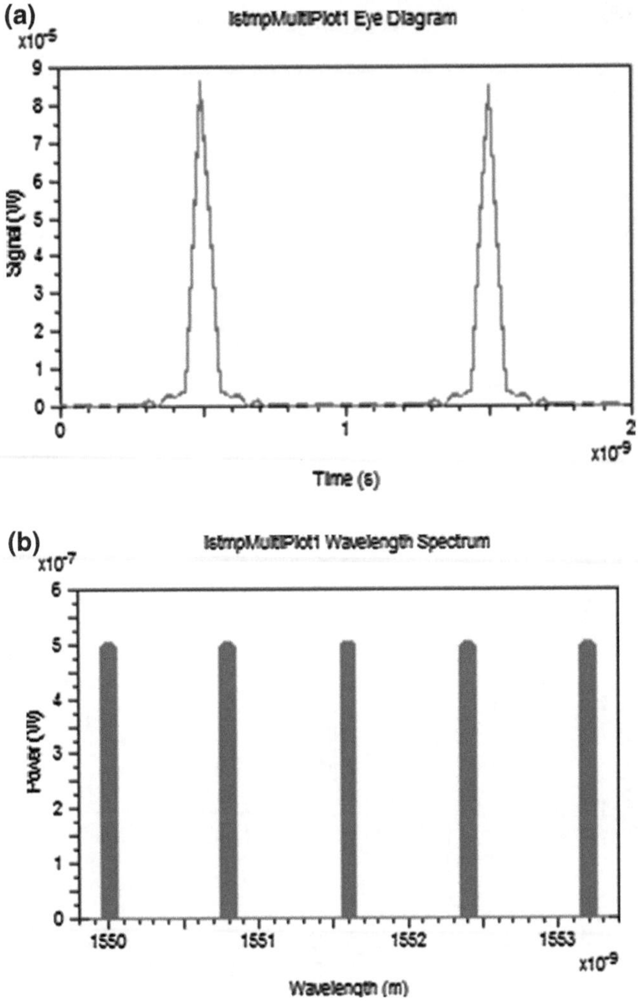

Fig. 8.34 **a** Eye diagram at input. **b** Signal strength at input at 1 Gbps

The main motto for the applications in communication systems is the voice, and for data networks, it is both the data and video. So to cater these applications, the development of the system has different requirements such as speed, symmetric or asymmetric access, and quality of service, so further, the applications are expanded as follows:

1. OCDMA-based fiber LAN and access networks (PON), e.g., FTTx, where x can be curb, home, office, etc.
2. Wireless OCDMA (indoor environment): Passenger plane multiaccess network, intra-spacecraft communication network, and communication between electronic subsystems in hospitals.

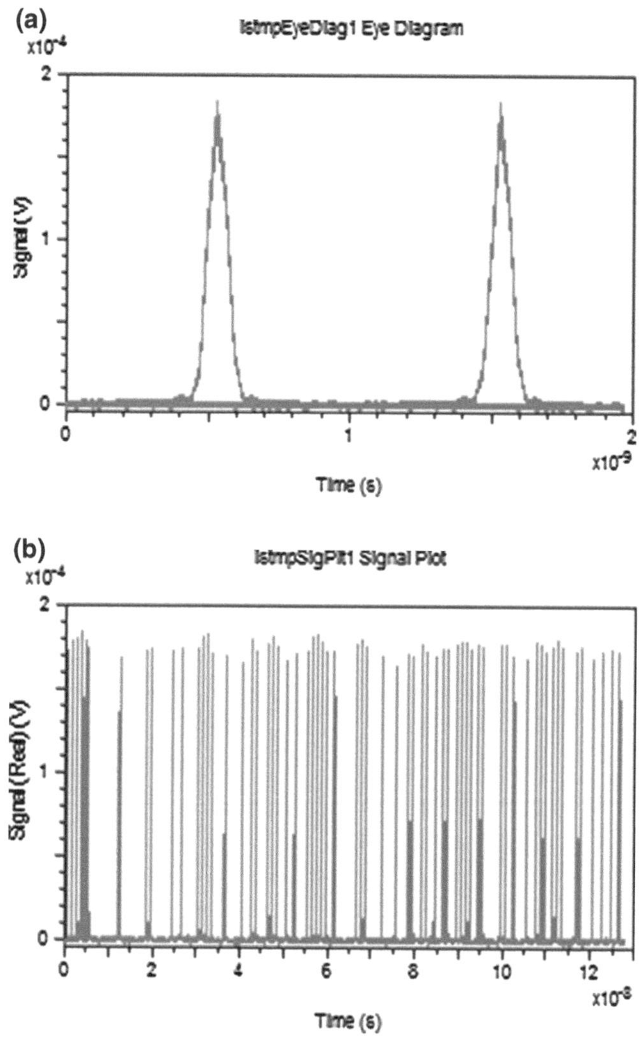

Fig. 8.35 **a** Eye diagram at −26 dBm. **b** Signal strength at −26 dBm at 1 Gbps

3. Metrolevel Optical VPN: The main goal of VPN (virtual private network) is to provide secure data links over insecure platform and optical VPN incorporating OCDMA codes provides enhanced security and due to inbuilt code structure, the signals can be decoded only at the corresponding end point.
4. Image Transmission: With OCDMA, during transmission and reception, image pixels are transmitted simultaneously over user specific code channels.
5. Radio over fiber: Due to high throughput, radio over fiber enables a fiber access network to connect wireless base stations. The application of OCDMA-based fiber network incorporates wireless sensor networks. Fiber radio networks can

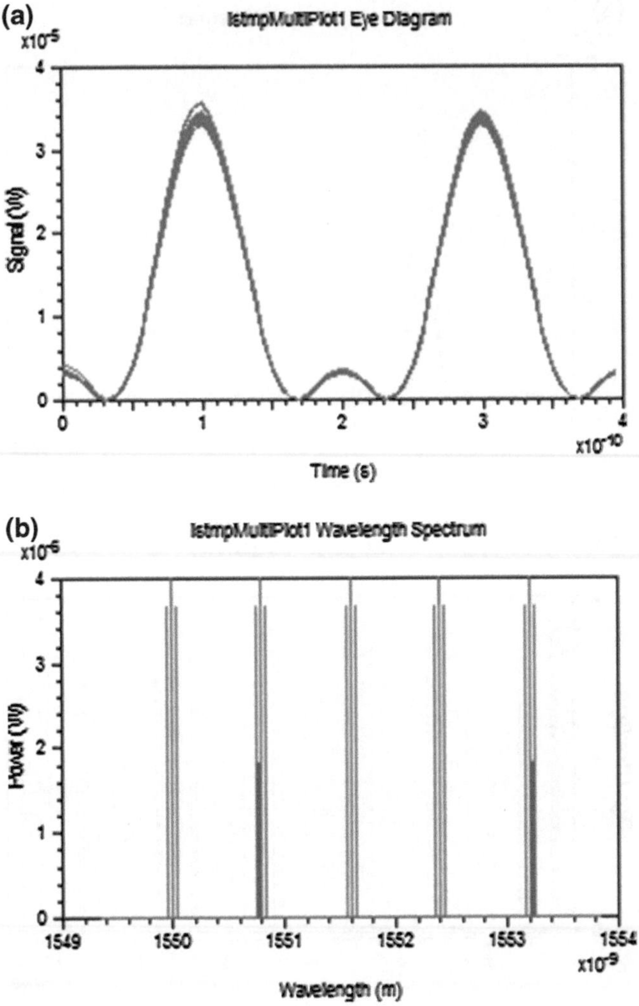

Fig. 8.36 **a** Eye diagram at input. **b** Signal strength at input at 5 Gbps

provide the communication between the sinks and the task manager node to simplify the design of sinks (radio-based stations).

8.14 Future of OCDMA

OCDMA is the technology of multiple access and networking due to in-built advantages such as random, simple management, flexible networking, supporting bursty traffic, good compatibility with other multiplexing technologies. With the

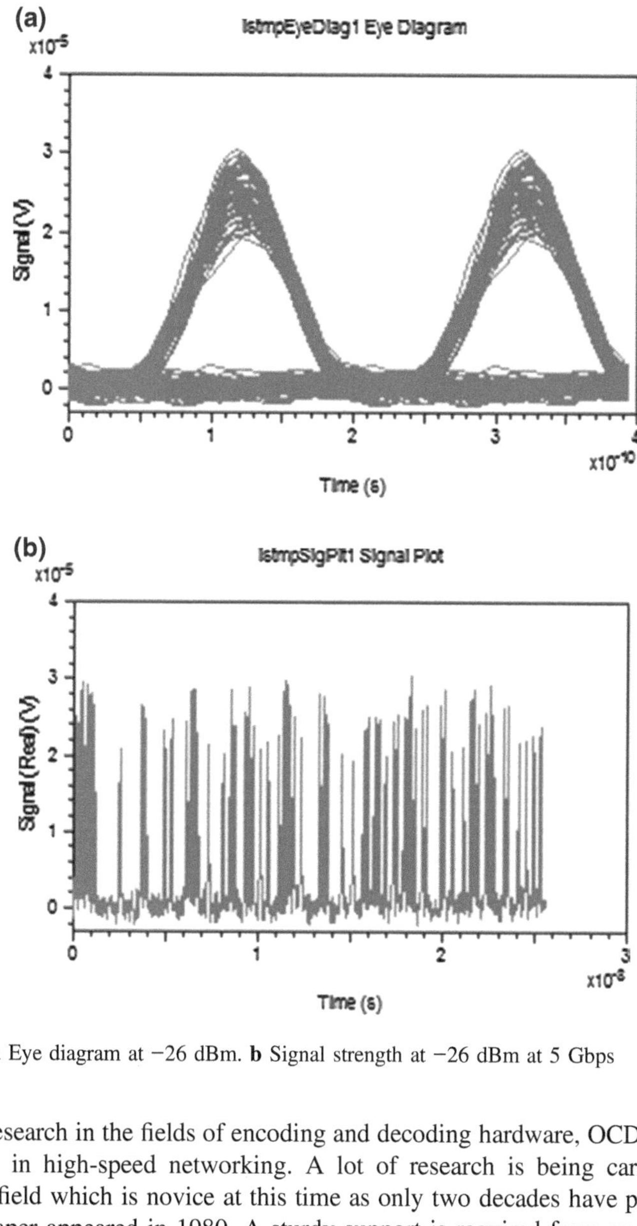

Fig. 8.37 **a** Eye diagram at −26 dBm. **b** Signal strength at −26 dBm at 5 Gbps

advance research in the fields of encoding and decoding hardware, OCDMA would find place in high-speed networking. A lot of research is being carried out in OCDMA field which is novice at this time as only two decades have passed since the first paper appeared in 1980. A sturdy support is required from research community, business venture, and policy makers to move the technology toward its practical implementation. With the stress for implementing OCDMA technology for future generation networks, it will find many application areas in the field of networking, multimedia, military, medical, communication, and entertainment industry.

References

1. ITU recommendations: G.694.2: spectral grids for WDM applications: CWDM wavelength grid
2. ITU Recommendations: G.694.1: Spectral grids for WDM applications: DWDM frequency grid
3. Chung, F.R.K., Salehi, J.A., Wei, V.K.: Optical orthogonal codes: design, analysis and applications. IEEE Trans. Inf. Theor. **35**(3), 595–604 (1989)
4. Salehi, J.A.: Code division multiple-access techniques in optical fiber networks-part I: fundamental principles. IEEE Trans. Commun. **37**(8), 824–833 (1989)
5. Yin, H.: Optical Code Division Multiple Access Communication Networks-Theory and Applications. Springer, Berlin (2009). ISBN 978-3-540-68445-9
6. Jindal, S., Gupta, N.: Comparison of spectral efficiency of CDMA communication system. Int. J. Commun. Eng. **2**(1), 29–32 (2010). ISSN No. 0975-6094
7. Shivaleela, E.S: Doctor of Philosophy Thesis Design and Performance Analysis of a New Family of wavelength/Time Codes for Fiber-Optic CDMA Networks. Indian Institute of Science Bangalore (2006)
8. Jindal, S., Gupta, N.: Member IEEE, simulated transmission analysis of 2D and 3D OOC for increasing the no. of potential users. 10th Anniversary International Conference on Transparent Optical Networks ICTON 2008 (organized by IEEE.) Athens, Greece, June 22–26 2008
9. Saghari, P., Kamath*, P., Arbab, V., Haghi, M., Willner, A.E., Bannister*, J.A., Touch, J.D.: Experimental demonstration of an interference-avoidance based protocol for O-CDMA networks. OFC, (2006). PDP 46
10. Rsoft's OPTSIM Models Reference Volume II Block Mode, 2010
11. Jindal, S., Gupta, N.: A new family of 3D code design using algebraic congruent operator for OCDMA systems. Int. J. Electron. Telecommun. Instrum. Eng. (IJETIE) **3**, 51–58 (2010). ISSN 0974-4975 (IF 2.)
12. Prucnal, P.R.: Optical Code Division Multiple Access: Fundamentals And Applications. CRC Press (2006)
13. Jindal, S., Gupta, N.: Member IEEE, analysis of 3D FO-CDMA system with varying data rates. National Conference on Emerging Trends in Engineering and Management (ETEM-2008), IEEE Student Branch Bombay Section, India. p. 6 (2008)
14. Jindal, S., Gupta, N.: Member IEEE, performance evaluation of optical CDMA based 3D code with increasing bit rate in local area network. IEEE Region 8, SIBIRCON-2008, International Conference on Computational Technologies in Electrical and Electronics Engineering Novosibirsk Scientific Centre, Novosibirsk, Russia, Sponsored by IEEE Region 8, Russia, 21–25 July 2008
15. Jaafar, H.B., LaRochelle, S., Cortès*, P.-Y., Fathallah, H.: 1.25 Gbit/s transmission of optical FFH-OCDMA signals over 80 km with 16 users. OFC, Tu V3-1 (2001)
16. Glesk, I., Bres, C.-S., Prucnal, P.R.: Demonstration of an eight-User 115-G chip/s incoherent OCDMA system using super continuum generation and optical time gating. IEEE Photonics Technol. Lett. **18**(7), 889–891 (2006). ISSN 1041-1135
17. Wang, X., Wada, N., Miyazaki, T., Cincotti, G., Kitayama, K.-I.: Field Trial of 3-WDM × 10-OCDMA × 10.71 Gbps, truly-asynchronous, WDM/DPSK-OCDMA using hybrid E/D without FEC and optical threshold. (2006). OFC, (2006). PDP 44
18. Cong, W., Scott, R.P., Hernandez, V.J., Kebin Li, B.H., Kolner, B. Heritage, J.P., Yoo, S.J.B.: Demonstration of a 6 × 10 Gb/s Multiuser Time-slotted SPECTS O-CDMA Network Testbed, Optical Society of America (2005)
19. Scott, R.P., Cong, W., Hernandez, V.J. Kebin Li, B.H. Kolner, B., Heritage, J.P., Yoo, S.J.B.: An Eight-User Time-Slotted SPECTS O-CDMA Test bed: Demonstration and Simulations. J. Lightwave Technol. **23**(10), 3232 (2005)

20. Hernandez, V.J.: 320-Gb/s capacity (32 users × 10 Gb/s) SPECTS O-CDMA local area network tested. OFC, pp. 1–3 (2006)
21. Jiang, Z., et al.: Four-user, 2.5-Gb/s, spectrally coded OCDMA system demonstration using low-power nonlinear processing. J. Lightwave Technol. **23**(1), 143–157 (2005)
22. Lu, L., et al.: Two-user 2.5 Gbps 100 km OCDMA transmission experiment using EPS-SSFBG en/decoder. In: Proceedings of SPIE-OSA-IEEE Asia Communications and Photonics, SPIE, vol. 7632, 76320 N (2009)
23. McGeehan, J.E., Motaghian Nezam, S.M. R., Saghari, P., Willner, A.E., Omrani, R., Vijay Kumar, P.: Experimental demonstration of OCDMA transmission using a three-dimensional (time-wavelength-polarization) codeset. J. Lightwave Technol. **23**(10), 3282–3289 doi: 10.1109/JLT.2005.856302 (2005)
24. Stallings, W.: Data and Computer Communications. 7th edn. Prentice Hall (2003)
25. Jindal, S., Gupta, N.: Analysis of multi dimensional codes for OCDMA system. CiiT Int. J. **4** (12), 732–737 (2012)
26. Omrani, R., Vijay Kumar, P.: Codes for Optical CDMA. In: Proc. of SETA. Springer-Verlags Lecture Notes in Computer Science Series, vol. 4086, pp. 34–46. Springer, Berlin, Heidelberg (2006)
27. Jindal, S., Gupta, N.: Exploration of three dimensional codes based on model A and model B using algebraic congruent operator in OCDMA system. Int. J. Mob. Adhoc Netw. **2**(3), 345–351 (2012)
28. Salehi, J.A.: Emerging optical CDMA techniques and applications. Int. J. Opt. Photonics **1**(1), 15–32 (2007)

Chapter 9
Focused Crawling: An Approach for URL Queue Optimization Using Link Score

Sunita Rawat

Abstract The hasty expansion of the World Wide Web poses exceptional scaling challenges for traditional crawlers and search engines. Web crawlers incessantly carry on crawling the Web and locate any novel Web pages that have been added to or removed from the Web. Because of dynamic and growing nature of the Web, it is tricky to deal with inappropriate pages and to forecast which links lead to excellence pages. Since the crawler is just a computer program, it cannot decide how pertinent a Web page is. In this paper, a method of efficient focused crawling is implemented to enhance the quality of Web navigation. We compute the unvisited URL score based on various factors such as its description in Google search engine and its anchor text relevancy and compute the similarity measure of description with given query or topic keywords. Relevancy score is calculated based on vector space model (VSM). Queue optimization is done on the basis of duplicate link and content similarity.

Keywords Focused crawler · Search engine · Weight table · Queue optimization

9.1 Introduction

With the enormous growth of information on the World Wide Web (WWW), there is a vast demand for developing effective and efficient techniques to systematize and retrieve the information available [1]. This information is frequently scattered among several Web servers and hosts, via various formats. Finding information of use from the WWW, which is a large-scale disseminated structure, need proficient search strategies. The majority of search engines keep an index of Web pages which is exploited to give up-to-date information associated with user-defined queries/ topic. Crawling/traversal of the entire Web takes weeks or some order of days, great

---

S. Rawat (✉)
Department of Computer Engineering, RCPIT, Shirpur, Dhule, India
e-mail: sajaldixit32@gmail.com

S. Patnaik et al. (eds.), *Recent Development in Wireless Sensor and Ad-hoc Networks*, Signals and Communication Technology, DOI 10.1007/978-81-322-2129-6_9

storage capacity, and network bandwidth. The delay in updation of index causes display of missing highly relevant pages to the user queries and/or irrelevant pages.

The WWW is growing more rapidly compared to network bandwidth, storage capacity, and processor speed, which needs efficient scalable solutions. To get information about a specific topic Web crawler goes through all the Web servers. Where searching all the pages and Web servers are not practical. As a result, numerous new ideas have been planned; amid them, a key method is focused crawling which aims to crawl specific topical portions of the Web rapidly without having to visit all Web pages [2]. Focused crawler is an automatic technique to efficiently find pages on the Web which are pertinent to a topic. Focused crawlers were proposed to navigate and fetch only a part of the Web that is pertinent to a specific topic, starting from seed pages, which are basically a set of pages from where crawling starts. It efficiently utilizes the storage capacity and network bandwidth. Focused crawling provides a feasible method for regular updation of indexes of search engine.

Focused crawlers [3, 4] plan to look for and retrieve only the subset of the WWW that pertains to a precise topic of relevance. The perfect focused crawler retrieves the maximal set of relevant pages while concurrently traversing the minimal number of irrelevant documents on the Web. Focused crawler cautiously decides in what order to follow and which URLs to scan depending on formerly downloaded pages information. Previous search engine which deployed the focused crawling approach based on the perception that pertinent pages often contain pertinent links [2]. The crawler is kept focused through a crawling approach which finds the relevancy of the Web page to the predefined topic keyword, and depending on this relevancy, a judgment is done for downloading of the Web page [5]. Focused crawler has been helpful for other appliances such as distributed processing of the whole web, with every one crawler allocated to a small part of the Web.

In this paper, we compute the link score. Initially, we compute the unvisited URL score depending on its description in Google search engine and its anchor text relevancy and compute the similarity measure of description with given query or topic keywords.

The rest of this paper is organized as follows. Section 9.2 discusses on related work of focused crawler. Section 9.3 describes the working of focused crawler. Section 9.4 discusses crawler design issues. Section 9.5 presents different crawling techniques. Section 9.6 discusses metrics used in crawling algorithms. Section 9.7 represents various approaches of crawling. Section 9.8 discusses about the URL normalization. Section 9.9 describes the architecture of focused crawler which we used. Section 9.10 presents our proposed approach. Section 9.11 represents our proposed algorithm, and after that, we have concluded our paper.

9.2 Related Work

Topic-oriented crawlers try to focus the crawling process on pages pertinent to the topic or query. They maintain the overall number of downloaded Web pages for processing [6, 7] to a least amount, while maximizing the percentage of pages which are pertinent. Their performance mainly depends on the selection of starting URLs or starting pages which are also known as seed pages or seed URLs. In general, users provide a set of seed pages as input to a crawler or, otherwise, seed pages are selected among the top answers returned by a Web search engine [8], considering the topic as query [9]. High-quality seed pages can be either pages pertinent to the topic or pages from which pertinent pages can be accessed within a few number of routing hops.

Crawlers used by general-purpose search engines get back huge numbers of Web pages in spite of their topic or query. Focused crawlers work by combining both the link structure of the Web and the content of the retrieved Web pages for assigning higher weightage to pages with higher chances of being pertinent to a given topic/ query. Focused crawlers can be categorized as follows:

(a) Classic Focused Crawlers:
Classic focused crawlers [10] get as input a user query that illustrates the topic, a set of starting page (seed) URLs and they direct the search in the direction of pages of interest. For assigning higher download priorities to links, they incorporate criteria based on their probability to lead to pages on the topic of query. Preference is given to pages which are pointed to by links with higher priority and are downloaded first. In this way, crawler proceeds recursively on the links contained in the downloaded pages. In general, download priorities are calculated depending on the similarity between the anchor text of a page link and topic, or between text of the page containing the link and the topic. In this case, text similarity is calculated by the vector space model (VSM) [11].

(b) Semantic Crawlers:
Semantic crawlers [12] are a deviation of classic focused crawlers. In this case, different criteria are used for deciding download priorities of pages. The topic and a page can be pertinent if they share conceptually (but not necessarily lexically) similar terms. Ontologies are used to define conceptual similarity between terms [12–13].

(c) Learning Crawlers:
Learning crawlers [14] use a training process for allocating visit priorities to Web pages and for directing the process of crawling. They are characterized by the way relevant Web pages or paths through Web links for reaching relevant pages are learned by the crawler. In general, a learning crawler is get trained by supplying it with a training set consisting of both relevant and not relevant kind of Web pages [15, 16]. Techniques based on hidden Markov models (HMM) [17] and context graphs [18] consider not only the page content and the corresponding classification of Web pages as pertinent or not pertinent to the topic but also the link structure of the Web and the likelihood

that a given page (which may be not pertinent to the topic) will lead to a pertinent page within a small number of steps (hops). By combining ideas from learning crawlers with ideas of classic focused crawlers, hybrid methods are form [19].

A focused crawler is a computer program used for finding information associated to some specific topic from the WWW. However, the main goal of focused crawling is that the crawler selects and retrieves pertinent pages only and does not need to gather all Web pages. As the crawler is only a computer program, it cannot predict how pertinent a Web page is [20]. In an attempt to search pages of a specific type or on a specific topic, focused crawlers aspire to recognize links that are probably to direct to target documents and pass up links to off topic. fish-search algorithm and shark-search algorithm were used previously for crawling with topic keywords mentioned in query.

In fish-search algorithm, the system is driven by the query. It starts from a set of seed pages; it takes only those pages that have content similar to a given query and their relative pages. The fish- search technique allocates binary priority values (0 for not relevant and 1 for relevant) to candidate pages for downloading using simple keyword matching. Hence, all pertinent pages are assigned the equal priority value. Shark-search is an advance algorithm of fish-search algorithm in which a child link inherits a partial value of the score of its parent link and this score is united with a value that depends on the anchor text that arises around the link in the Web page [21]. The shark-search method proposes using VSM for allocating non-binary priority values to candidate pages. Other methods to focused crawling include best-first crawler and InfoSpiders [22]. Best-first crawlers allocate priority values to candidate pages by calculating their text similarity with the topic by using VSM [11], while InfoSpiders use neural networks. Shark-search can be seen as a deviation of Best-first crawler with an extra complicated priority assignment function. Best-first crawlers have been shown to do better than shark-search and InfoSpiders, as well as other non-focused breadth-first crawling methods. Due to its simplicity and effectiveness, Best-first crawling is supposed to be the most successful approach to focused crawling. An algorithm founded on ontology considers in [15] for calculation of page relevance. Next to preprocessing, entities are took out from the page and counted. Relevancy of the page with respect to user-selected query is then calculated by using numerous measures on ontology graph. The harvest rate is enhanced compared to the baseline focused crawler.

A focused crawling process consists of two important stages. The first stage is to decide the starting URLs and identify user interest. (Because the crawler is incapable to traverse the Web without starting URLs). The second stage is the crawling technique. A focused crawler elegantly chooses a direction to traverse the Internet. Best route selection technique of the focused crawler is to position URLs in such a way that the most pertinent ones can be situated in the first part of the queue. Depending on the relevancy of the URLs, queue will then be sorted [16]. The efficiency and performance of a focused crawler is primarily decided by the ordering strategy that decides the order of page retrieval.

9.3 Working of Web Crawler

Essential components to search engines are Web crawlers; running a Web crawler is a tough job. Crawling is the most delicate application as it involves working together with hundreds of thousands of Web servers, due to which control of the system is very hectic. Speed of Web crawling is governed by two factors: First is speed of sites that are to be crawled and second one is speed Internet connection [23].

Following is the procedure by which Web crawlers work:

- Download the Web page.
- Go through the downloaded page and get back all the links.
- Repeat the procedure, for each link retrieved.

Figure 9.1 shows the flow chart of a basic sequential crawler, in which working of crawler is shown in detail.

For crawling through an entire site on the Internet/intranet, the Web crawler can be used. You give a start URL, and the Crawler follows all links available in that HTML page. It generally directs to further links again and again. Therefore, any site can be seen as a tree structure, where the root is the start URL; all links in that root HTML page are direct sons of the root. Succeeding links are then sons of the previous sons. As we know, crawler exists on a single machine. When user clicks on links, Web browser sends HTTP request to other machines on Internet, in the same way the crawler sends HTTP requests for documents. Web crawling can be look upon as processing items in a queue. Whenever the crawler visits a Web page,

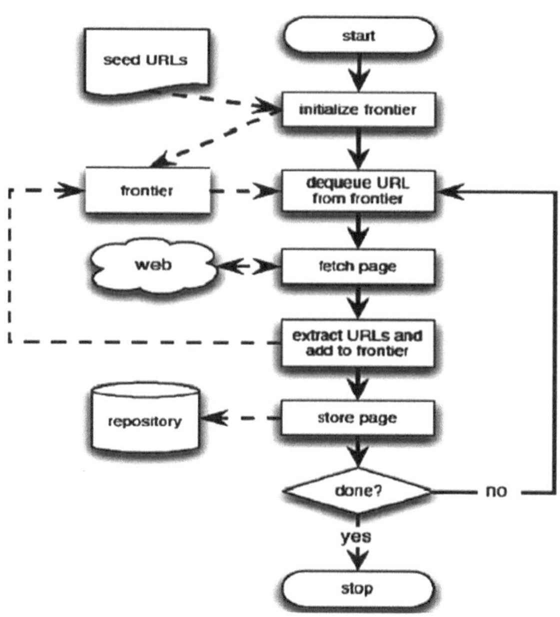

Fig. 9.1 Flow chart of a basic sequential crawler [23]

it takes out links to further Web pages. Therefore, the crawler places these URLs at the end of a queue and carries on crawling to a URL that it removes from the front of the queue [23].

9.4 Crawler Design Issues

Issues associated to crawler design are discussed as follows:

(A) Input:
In the case of focused crawlers, it takes as input a number of starting (seed) URLs and the topic description. Topic description can be a training set for learning crawlers or a query given by the user or list of keywords for classic and semantic focused crawlers.

(B) Page Downloading:
All the links present in downloaded pages are extracted and positioned in a queue. Traditional crawler uses these links and continues with downloading new pages in a first-in, first-out (FIFO) manner. On the other hand, focused crawler works, on the basis of content relevancy, and it rearranges queue entries or might decide to eliminate a link from further expansion.

(C) Content Processing:
Relevancy of a page is calculated by applying term frequency–inverse document frequency vector (TF-IDF) according to VSM. Downloaded pages are lexically analyzed by applying a stemming algorithm [7], and stop words are removed and reduced into term vectors. Most Best-first crawler implementations use only term frequency (TF) weights because as per assumptions, computing inverse document frequency (IDF) weights during crawling can be difficult.

(D) Priority Assignment:
The URLs which are extracted from downloaded pages are positioned in a priority queue where priorities are decided based on the relevancy of page to a query topic.

(E) Expansion:
From the downloaded pages, URLs are chosen for more expansion and steps (B)–(E) are repeated until some conditions (such as the required number of pages has been downloaded) are fulfilled or system resources are exhausted.

9.5 Crawling Techniques

(A) Distributed Crawling:
Due to the growing and dynamic nature of the Web, indexing of it is a very difficult job. Day by day, size of the Web is growing, and it has become crucial

to weak the crawling process so as to end downloading the pages in a reasonable amount of time. When a single centralized crawler is used, all the retrieved data go through a single physical link. When crawling activity is get distributed via multiple processes, it can help build an easily configurable, scalable system, which is fault tolerant system. Distributing the load reduces hardware requirements and simultaneously improves the overall reliability and download speed. No central coordinator exists, that is, each task is performed in a fully dispersed way [24].

(B) Focused Crawling:
A traditional Web crawler collects as many pages as it can from a specific set of URLs, while a focused crawler is designed to collect documents on a particular topic, therefore reducing the amount of downloads and network traffic. The objective of the focused crawler is to selectively find out pages that are pertinent to a predefined set of topics or query.

(i) Architecture of Focused Crawler:
The focused crawler is made up of four sub-systems:

- Fetching of seed pages
- Generation of topic keywords
- Similarity measurement
- Spider

Figure 9.2 shows the entire working process of the focused crawler.

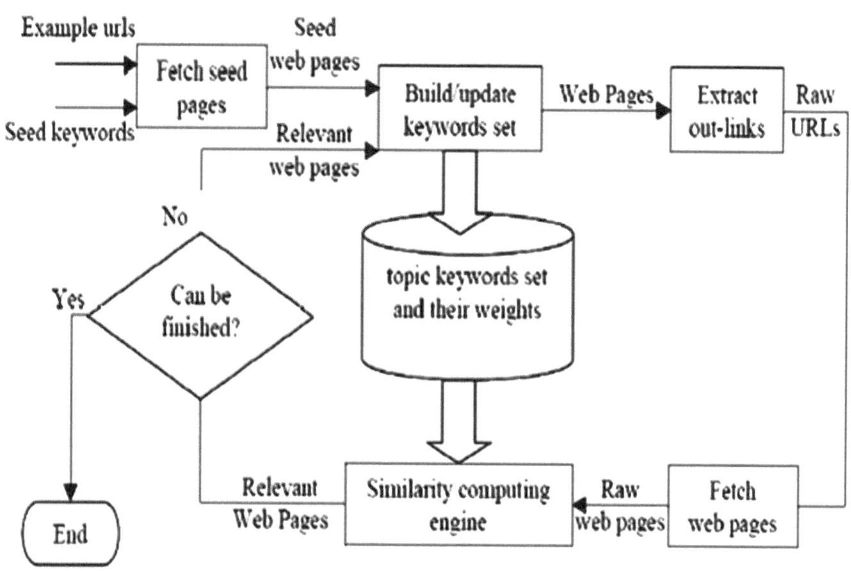

Fig. 9.2 Focused crawler working process [30]

- Fetching of Seed Pages:

Given a set of topic keywords, first of all, system searches them on a search engine such as Google. Search engine would return huge set of result. The top *N* URLs are mostly pertinent to the topic. These top *N* URLs are used by the crawler as seed URLs to fetch seed pages from Web. Figure 9.2 shows how the focussed crawler generates the seed pages.

- Generation of Topic Keywords

For user, it is tough to specify all keywords associated with the topic. Suppose if large amount of documents which are pertinent to the topic are given, then it is easy to find out most significant keywords to represent the topic. These keywords are known as topic keywords. Aim of this sub-system is to find topic keywords. Term frequency (TF_i) is get calculated for each word W_i in document, first. Then, inverse document frequency IDF is get calculated. Depending on this TF-IDF value, the importance of a word to a document is get decided. Finally, top *N* uppermost weight keywords are treated as topic keywords set.

- Similarity Measurement

When a new Web page is fetched, the crawler's job is to check whether the page is pertinent to the topic or not. Suppose the document *D* is a Web page that required to be checked. The query *Q* represents a set of topic keywords. Similarity measurement sim(*Q*, *D*) gives a float value which ranges in between 0 and 1. Let θ be the threshold value to judge the relevancy of the document. If the value of θ is higher, the accuracy of retrieved pages pertinent to the topic would also be higher.

- Spider:

The job of spider is very simple. It takes set of URLs as input, and it outputs the equivalent set of Web pages.

9.6 Metrics Used in Crawl Algorithms

(A) Cosine Similarity Measure
The most important stage for focused crawling is analysis of topic similarity. After the collection of Web pages, the job of crawler is to decide the relevancy of the page. As the crawler cannot decide which Web page is pertinent to the topic, a similarity measure should be used to decide the relevancy of a Web page. Various techniques are used to compute the topic similarity. One of them is a VSM [11]. In VSM, a Web page is represented as a vector. In that page, each dissimilar word is considered to be an axis of a vector in a

multi-dimensional space. Two Web pages are considered to be pertinent, i.e., concerning to the same topic, scientifically, two vectors point to the same direction indicates that they have an angle of zero degree between them and their cosine rate becomes 1. Hence, a vector matching function, based on the cosine correlation, is used to calculate the degree of topic similarity between the interested topic and the page.

(B) Term Weighting

According to a given weighting model, terms are weighted which might include local (i.e., at the level of documents) weighting, global (i.e., at the level of database collections) weighting, or both. When local weights are used, the weight of a term is given by term frequencies, *tf*. When we consider global weights, at that time, term weights are defined as inverse document frequencies *idf* values. Very popular weighting scheme is one in which both local and global weights are used (weight of a term = *tf* * *idf*). Generally, it is referred as *tf* * *idf* weighting. This weighting scheme is expressed in detail in proposed approach.

9.7 Various Approaches of Crawling

Two main approaches used for crawling are:

- Breadth-First Approach
- Best-First Approach

(A) Breadth-First Approach

This is the easiest strategy used for crawling. For deciding which URL to visit next, it does not utilize heuristics. All URLs in the current level will be visited in the same order in which they are found before next-level URLs are visited. To build domain-specific collections, breadth-first search could be used. Here, there is one assumption that if the starting URLs are pertinent to the target topic, there are chances that pages in the next level are also pertinent to the target topic. However, the size of collections made by such crawlers cannot be large because after a particular limit, breadth-first search starts to lose its focus and collects lot of irrelevant pages.

Figure 9.3 illustrates the breadth-first crawler. Usually, breadth-first algorithm is used as a baseline crawler; as it acts blindly, it does not use any knowledge about the topic. Therefore, it is also called blind search algorithm. Performance of this algorithm is used to provide a lower bound for any of the more sophisticated algorithms.

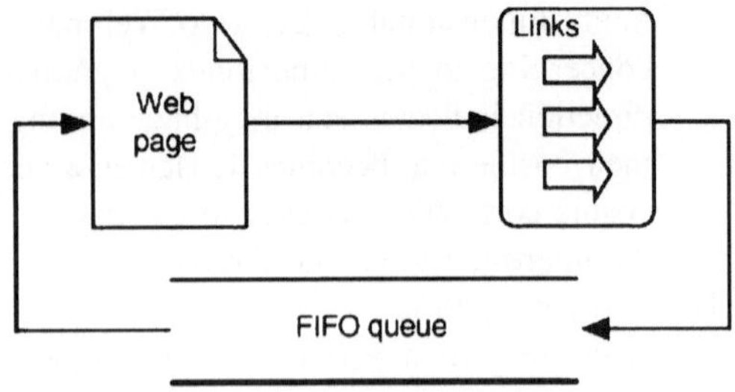

Fig. 9.3 Breadth-first crawling approach [20]

- Limitations of Breadth-First Approach

There are millions of pages, in real WWW structure, linked to each other. As the size of the repository of any search engine is restricted, it cannot accommodate all pages. Therefore, it is desired that only most suitable and relevant pages we store in repository. Disadvantage of this algorithm is that URLs traversing is done in sequential order. When the total number of pages is small, it may be good. However, in real life, a lot of useless pages can generate links to other useless pages. Therefore, storing and processing such kind of links in frontier is wastage of memory and speed. Every time, for processing, we should select a useful page from the frontier, irrespective of its position in the frontier. As breadth-first technique traverses in sequential order, it is not desirable.

(B) Best-First Approach:

Cho et al. [4] studied a heuristic approach called best-first crawling approach to overcome the problems of breadth-first approach. In this approach, next link for crawling is selected from a given frontier of links, on the basis of some priority or score. Therefore, always the best available link is opened and traversed. By using various mathematical formulas, priority or score of each link can be computed. Figure 9.4 illustrates the best-first crawler.

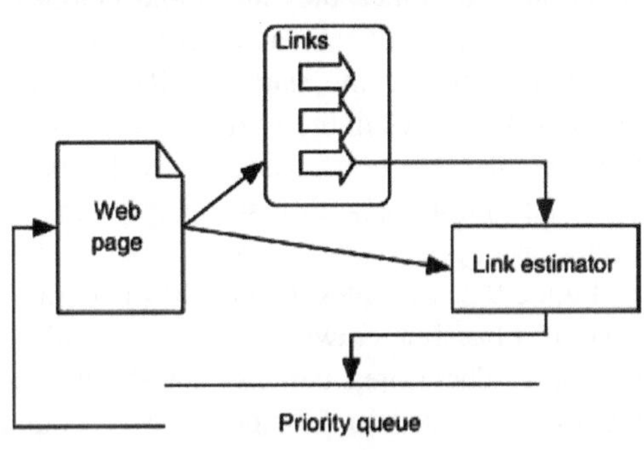

Fig. 9.4 Best-first crawling approach [20]

Following Web Crawling Algorithms use Heuristic Approach:

(i) Naive Best-First Algorithm

A naive best first was one of the crawlers studied by the authors [11], for crawler evaluation. Fetched Web page is represented as a vector of words by this crawler. Where weights of words are calculated by its occurrence frequency. Afterward, the crawler calculates the cosine similarity of the page to the given topic or query given by the user. And depending on this similarity value, the unvisited URLs on the page are then added to a frontier which is maintained as a priority queue. The cosine similarity between the page p and a query q is computed by

$$\sim (q,p) = \frac{V_q.V_p}{||V_q||.||V_p||}$$

where V_q and V_p are vector representations of the query and the page, respectively, V_q. V_p is the dot product of the two vectors, and $\|v\|$ is the Euclidean norm of the vector v.

(ii) PageRank

PageRank value is calculated off-line for each page, and there is no dependency on search queries. Hence, PageRank is a static ranking of Web pages. Basically, PageRank represents a hyperlink from page x to page y as a vote. PageRank is nothing but number of links or votes that a page receives. As the number of votes received by page increases, its importance also get increases.

- Strengths and Weaknesses of PageRank:

The main benefit of PageRank is its capability to fight spam. Basically, a page is significant if the pages pointing to it are significant. As it is difficult for Web page owner to insert in-links into her/his page from other significant pages, it is thus difficult to manipulate PageRank. However, there are some methods to manipulate PageRank. Identifying and struggling spam is an important matter in Web search.

Second most important advantage of PageRank is that it is a query independent and global measure. Namely, the PageRank values of all the pages on the Web are calculated and saved off-line instead of at the query time [23]. There is great contribution of both of these two advantages to Google's success.

In recent times, PageRank has been used to improve page quality by guiding the crawlers. As PageRank calculates the importance of a Web page, it must be used for ranking the results of a search by combining it with one or more measures of query relevance.

(iii) HITS Algorithm:

HITS stands for hypertext induced topic search [23]. HITS is search query dependent, unlike PageRank which is a static ranking algorithm. When the user inputs a search query, the list of relevant pages returned by a search engine is first

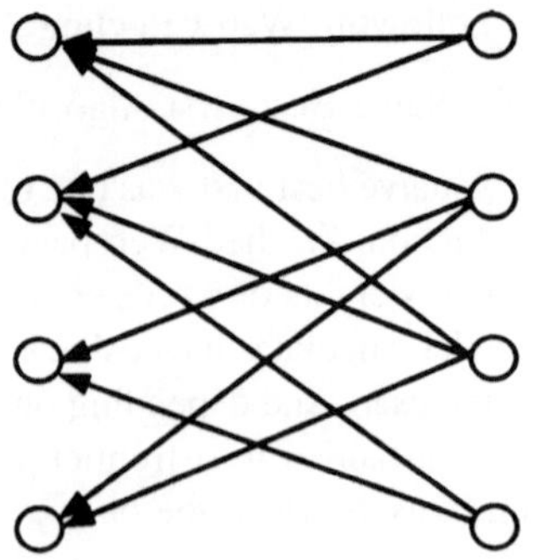

Fig. 9.5 A densely linked set of authorities and hubs [23]

get expanded by HITS and then produces two rankings of the expanded set of pages, hub ranking and authority ranking.

An authority is a page with many in-links. Where concept is that the page might have good quality or authoritative content on some topic, and therefore, many people believe it and link to it. However, a hub is a page with many out-links. Here, idea is that the page serves as a coordinator of the information on a specific topic and points to many good influence pages on the topic. So whenever a user comes to this hub page, he/she will get numerous useful links which guide him/her to good content pages on the topic. Hence, main idea of HITS is that a good hub points to many good authorities and a good authority is pointed to by many good hubs. Therefore, authorities and hubs have a mutual reinforcement link. A set of densely linked authorities and hubs are shown in Fig. 9.5.

- Strengths and Weaknesses of HITS:

The ability to rank pages according to the query topic is the main strength of HITS [23], which may be able to provide more pertinent authority and hub pages.

However, HITS has numerous disadvantages.

Firstly, there is absence of the anti-spam facility of PageRank. To influence HITS is quite easy by inserting out-links from one's own page to point to many good authorities. Since authority and hub scores are interdependent, when hub score increases, it in turn also increases the authority score of the page.

Second problem of HITS is topic drift. While expanding the root set, it can effortlessly collect many pages which are irrelevant to the search topic because out-links of a page may not point to pages that are pertinent to the topic and in-links to pages in the root set might be unrelated as well, because people put hyperlinks for all kinds of reasons, including spamming.

One more drawback is the query time evaluation. Getting the root set, expanding it, and then performing eigenvector calculation are all time-consuming operations.

9.8 URL Normalization

During the process of Web crawling, the tasks perform by the Web crawler is standard URL normalization. For converting URLs into canonical arrangement, there is a set of predefined procedures to be done. Crawling of redundant Web pages can be reduced by normalization, as URLs which are syntactically matching are assumed as similar [25].

Figure 9.6 shows the URL components.

To extend the standard URL normalization, various research works have been done [26, 27]. Objective of all these is to identify similar URLs syntactically. There is possibility that syntactically different URLs lead to alike Web pages. Redirected Web pages, mirrored pages, and many others are the reason which could lead to such kind of situation.

Following is the discussion about the standard URL normalization, which can be divided into normalization based on syntax, normalization based on scheme, and normalization based on protocol [28].

Standard URL Normalization

(a) Normalization based on syntax:

- Case normalization: Translate all authority components and letters at scheme to lower case.
- Path segment normalization: From the path component, eliminate all dot segments, such as "." and "..".

(b) Scheme-based Normalization:

- Insert backslash "/" following the authority component of URL.
- Eliminate default port number, such as http scheme port has number 80.
- Shorten the fragment of URL.

(c) Protocol-based Normalization:

- When the outcome of accessing the resources is similar, then only it is appropriate.

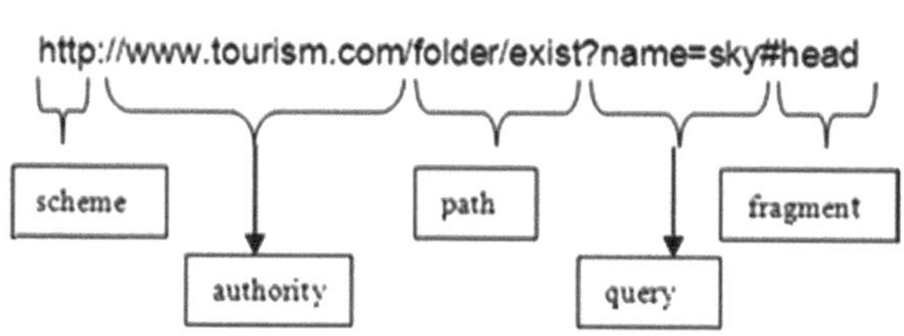

Fig. 9.6 URL components

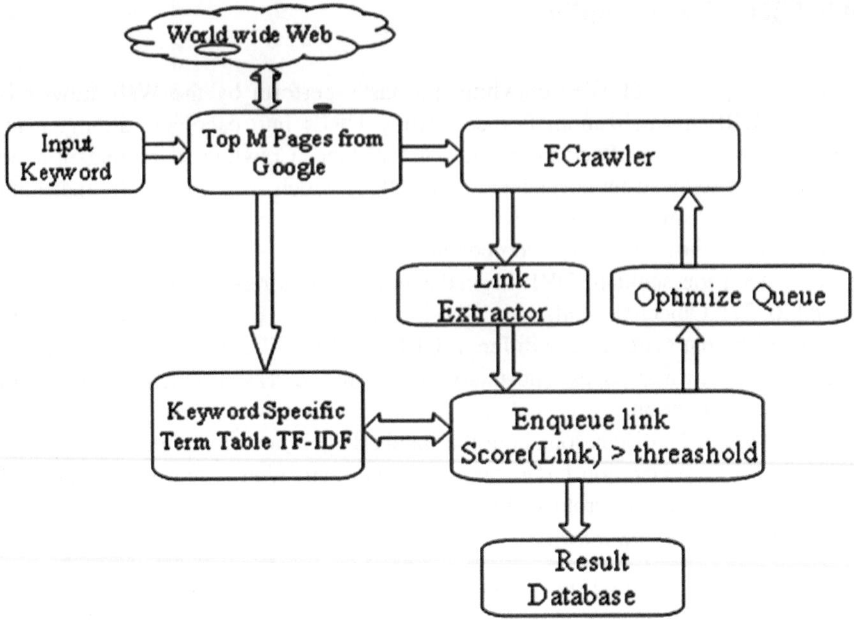

Fig. 9.7 System architecture

9.9 System Architecture

Figure 9.7 shows the proposed architecture. Depending on the input keyword or query given by the user, Google search engine extract pages from WWW. From those resultant pages, top "m" pages are act as seed URLs. TF-IDF calculates importance of a word to a document in a corpus. By using TF-IDF, keyword-specific term table is created. The link extractor extracts information such as hyperlink URLs from a downloaded page. Relevancy calculator calculates significance of a page regarding the topic and allocates score to URLs extracted from the page. If link score is greater than threshold value, then it will be added to the URL queue, or else URLs will be ignored.

9.10 Proposed Approach

(A) Extraction of Seed URL
In this approach, for extracting seed URLs, we have used Google search engine. When we place a query in this Google search engine, it shows the results of that query. We take first "*m*" resulting URLs as a seed URLs. We consider here only Google search engine for finding seed URLs, as no other search engine is as

powerful as Google. We suppose that those "*m*" resultant URLs are more relevant for this query, and therefore, these URLs are the seed URLs.

(B) Construction of Topic-Specific Weight Table
Weight table expresses the crawling goal. Google search engine generates seed URLs for the topic name send as a query. Parsing is done on these retrieved pages. To avoid indexing ineffective words, a system known as a text retrieval system is used, which often links a stop list with a set of documents. A stop list is nothing but a set of words that are supposed "irrelevant." Such as, the, for, of, with, a, and so on are stop words, although they may appear regularly. Porter stemming algorithm is used for stemming the words. For instance, the collection of words play, played, playing, share a common word stem, play, and all are various occurrences of the same word play.

TF-IDF is a mathematical statistic. It calculates importance of a word to a document in a corpus.

1. *Term Frequency* (*TF*): A weight is assigned to each term in a document depending on the number of time that term occurs in the document. This weight is referred to as TF.

$$tf(t,d) = \text{freq}(t,d) \tag{9.1}$$

where

tf(*t*, *d*)	frequency of term *t* in document *d*
freq(*t*, *d*)	number of *occurrences* of *t* in *d*

2. *Inverse Document Frequency* (*IDF*): TF suffers from a significant problem, and all terms are considered equally important when it comes to assessing relevance on a query. In reality, certain terms have little power in characterizing the document like stop words.

$$idf(t,d) = \log(|D|/|d \epsilon D{:}t \epsilon d|) \tag{9.2}$$

where

$|D|$: Total number of documents in the corpus,

$|d \in D{:}\ t \in d|$: Number of documents in which the term appears, i.e., $tf(t, d) \neq 0$.

Suppose, if the term is not present in the collection, it will create a division by zero. As a result, change the formula to

$$1 + |d \epsilon D{:}t \epsilon d|$$

Table 9.1 Topic-specific weight table

Terms	Weight
Cricket	1.0
Star	0.9230769230769231
Deal	0.7179487179487181
Article	0.6666666666666667
Score	0.5897435897435898
Target	0.5641025641025641
England	0.5641025641025641
Capton	0.5641025641025641
India	0.5128205128205128
Test	0.46153846153846156

Then, *tf–idf* is calculated as:

$$tfidf(t,d,D) = tf(t,d) * idf(t,D) \tag{9.3}$$

Order the words depending on their weights and take out a certain number of words having high weight as the topic keywords. Then, normalization of weights is done like:

$$W = \frac{Wt_i}{Wt_{\max}} \tag{9.4}$$

where "Wt_i" and "$Wt_{\max}$" are the weights of keyword "i" and keyword with highest weight, respectively [29].

For instance, we have taken a topic keyword "cricket." For construction of topic-specific weight table, we place the "cricket" as a query in the Google Web search engine and retrieved the first 5 results. Then, remove stop words and stem the words. After that TF-IDF of each word is calculated for calculating the weights. At this time, we have considered top 10 most occurrences words. Here, the weight is normalized by using the Eq. (9.1) and topic-specific weight table is produced as shown in Table 9.1.

(C) Calculation of Relevancy of Page

The exact weight of words in page matching to the keyword in the topic-specific weight table is deliberated. Here, we have used the same technique for weight calculation as we have used for topic-specific weight table. In our proposed approach, to calculate the relevancy of the page with respect to a particular topic, it uses a cosine similarity measure.

$$\text{Relevance}(t_{wt} \cdot p_w) = \frac{\sum k\epsilon(t \cap p) wkt\, wkp}{\sqrt{\sum k\epsilon\, t(wkt)^2 \sum k\, \varepsilon\, t(wkp)^2}} \tag{9.5}$$

Here, "t_{wt}" is the topic-specific weight table, "p_w" is the Web page under examination, and "*wkt*" and "*wkp*" are the weights of keyword "*k*" in the weight table and in the Web page correspondingly.

(D) Calculation of Link Score

From experimentation, it has been confirmed that the seed pages are most relevant to given topics. As the focused crawler finished with fetching of seed pages from WWW, next step is to calculate link score. In our proposed approach, the link score is basically used to assign scores to unvisited links which are take out from the downloaded page. The link score is assigned based on the metadata of hyperlinks and the information of pages that have been crawled.

$$\begin{aligned} \text{Link\_Score}(i) = {} & \text{Anchor\_Relevancy\_Score}(i) \\ & + \text{Relevance\_Score\_URL\_Description}(i) \\ & + [\text{Relevancy}(P_1) + \text{Relevancy}(P_2) \\ & + \cdots + \text{Relevancy}(P_n)] \end{aligned}$$

where Anchor_Relevancy_Score is the relevancy score among anchor text and topic keywords. Our proposed approach computes Anchor_Relevancy_Score as anchor text depicts some information concerning to URL. Relevancy_Score URL description is the relevancy score of URL description regarding topic keywords. Table 9.2 shows the weight of topic keywords in description of URLs.

Table 9.2 Relevancy_Score_Url_Description table

Terms	Weight
Cricket	0.8434587345691274
Star	0.7347912239112454
Deal	0.7347912295375322
Article	0.7347912295375322
Score	0.7347912295375322
Target	0
England	0.9633677823245663
Capton	0.9509568234876904
India	0.9489234987983456
Test	0.9356432121234560

(E) Queue Optimization

The focused crawler's efficiency is not only based on crawling maximum number of pertinent pages, but it also based on time required to complete crawling process. For better efficiency, the time should be as least as possible. So to minimize operation time, we can focused on how to reduce the number of crawling step, and the number of crawling step can only be reduced by decreasing the number of link required to process by focused crawler. But problem is that there are so many links which have relevant link score and because of that, they get added into the URL queue. As every link in queue has relevant link score, we cannot remove any link from queue without any reason. So the solution for this problem is to optimize the queue. We can use following steps for URL queue optimization.

(a) To look for duplicate link
In a Web, it is very common to have multiple sources for a single page. So by using any source of them, we will reach to the same page. Therefore, while crawling, it is very much possible that a single link can occur in different number of pages, because of that, the same link can be extracted multiple times and copied into the URL queue. We can avoid processing of same link again and again by removing duplicate links.
(b) To look for content similarity
There is a possibility that the multiple Web pages with different names sharing the same exact copy of information, we can avoid the processing of same data by checking the content with previously processed Web pages.
(F) Success Ratio Table
The comparison between conventional crawler and focused crawler can also be represented with success ratio. The success ratio is the number of unwanted links saved by focused crawler over conventional crawler.

For example, in Table 9.3, for a query "cricket," the total number of links obtained by the crawler after crawling 5 pages, which are nothing but the seed pages, are 1,233. And the links selected by focused crawler are 715, so the success ratio can be calculated as:

Table 9.3 Success ratio of focused Crawler

Topic	Max_url	Total link	Link selected by FC	Ratio (%)
Cricket	5	1,233	715	42
Shahrukh Khan	5	452	301	33
Agriculture	5	1,258	385	67
Computer	5	580	257	55
Politics	5	615	195	68

For total 1,233 links, it saved

$$1,233 - 715 = 518\,\text{links}$$

To get the result in percentage→
Consider

$$\text{for } 1,233\,\text{links it saved} \;\rightarrow 518 \text{ links}$$

So

$$\text{for } 100\,\text{links how much it saved?}$$
$$100\,\text{links} \rightarrow ?$$
$$((1,233 - 715) * 100)/1,233 = 42\,\%$$

Hence, for query "cricket," the focused crawler with queue optimization saved 42 % of links from processing.

9.11 Proposed Algorithm

Step 1: Input keyword and maximum URL to crawl = MaxURL
Step 2: Retrieve top *M* pages from Google and enqueue first link in queue
Step 3: Create topic-specific term table using TF-IDF
Step 4: Dequeue link from queue and retrieve respective page
Step 5: Extract all link from page
Step 6: Assign score to all extracted link based on

  (a) Title relevancy score,
  (b) Relevancy score URL description
  (c) Anchor text relevancy score

Step 7: Rank link based on link score and remove all bottom non-relevant link
Step 8: Enqueue selected link in queue
Step 9: Optimize queue based on

  (a) Duplicate link
  (b) Content similarity

Step 10: If size of queue = MaxURL + MaxURL * 0.4, then stop crawling.

9.12 Conclusion and Future Work

A focused crawler is a Web crawler that aspires to download Web pages that are pertinent to a predefined query or topic. To find out whether the Web page is related to a particular topic, our focused crawler used link score technique. Link scoring decides which links to be crawled and assigns priority among them. The accurate selection of the pertinent links provides performance enhancement in both page relevance and execution time.

In this paper, focused crawler is implemented by using various link score techniques and queue optimization techniques. Queue is optimized based on duplicate link and content similarity. The appliance of these techniques provides substantial performance improvement. As a future work, we plan to apply different queue optimization technique and semantic approaches for focused Web crawling.

References

1. Pal, A., Tomar, D.S., Shrivastava, S.C.: Effective focused Crawling based on content and link structure analysis. (IJCSIS) Int. J. Comput. Sci. Inf. Sec. **2**(1) (2009)
2. Hati, D., Sahoo, B., Kumar, A.: Adaptive focused Crawling based on link analysis. In: 2nd International Conference on Education Technology and Computer (ICETC) (2010)
3. Chakrabarti, S., van der Berg, M., Dom, B.: Focused Crawling: a new approach to topic-specific web resource discovery. In: Proceedings of the 8th International World-Wide Web Conference (WWW8) (1999)
4. Cho, J., Garcia-Molina, H., Page, L.: Efficient Crawling through URL ordering. In: Proceedings of the Seventh World-Wide Web Conference (1998)
5. Cheng, Q., Beizhan, W., Pianpian, W.: Efficient focused Crawling strategy using combination of link structure and content similarity. IEEE (2008)
6. Pivk, A., Cimiano, P., Sure, Y., Gams, M., Rajkovic, V., Studer, R.: Transforming arbitrary tables into logical form with TARTAR. Data Knowl. Eng. **60**(3), 567–595 (2007)
7. Chang, C., Kayed, M., Girgis, M.R., Shaalan, K.F.: A survey of web information extraction systems. IEEE Trans. Knowl. Data Eng. TKDE-0475-1104.R3 (2006)
8. McCown, F., Nelson, M.: Agreeing to disagree: search engines and their public interfaces. In: ACM IEEE Joint Conference on Digital Libraries (JCDL 2007), pp. 309–318. Vancouver, British Columbia, Canada, 17–23 June 2007
9. Bao, S., Li, R., Yu, Y., Cao, Y.: Competitor Mining with the web knowledge. IEEE Trans. Data Eng. **20**(10), 1297–1310 (2008)
10. Menczer, F., Pant, G., Srinivasan, P.: Topical web Crawlers: evaluating adaptive algorithms. ACM Trans. Internet Technol. (TOIT) **4**(4), 378–419 (2004)
11. Salton, G., Wong, A., Yang, C.S.: A vector space model for automatic indexing. Commun. ACM **18**(11), 613–620 (1975)
12. Ehrig, M., Maedche, A.: Ontology-focused Crawling of web documents. In: Proceedings of the Symposium on Applied Computing (SAC 2003), 9–12 Mar 2003
13. Hliaoutakis, A., Varelas, G., Voutsakis, E., Petrakis, E.G.M., Milios, E.: Information retrieval by semantic similarity. Int. J. Seman. Web Inf. Syst. (IJSWIS) Spec. Issue Multimedia **3**(3), 55–73 (2006)
14. Pant, G., Srinivasan, P.: Learning to Crawl: comparing classification schemes. ACM Trans. Inf. Syst. (TOIS) **23**(4), 430–462 (2005)

15. Ehrig, M., Maedche, A.: Ontology-focused Crawling of web documents. In: Proceedings of the Symposium on Applied Computing, March, 67.2–12.72 Florida, USA (2003)
16. Yuvarani, M., Ch., N., Iyengar, S.N., Kannan, A., Crawler, L.S.: A framework for an enhanced focused web Crawler based on link semantics. In: Proceedings of the IEEEIWIC/ ACM International Conference on Web Intelligence (2006)
17. Liu, H., Janssen, J., Milios, E.: Using HMM to learn user browsing patterns for focused web Crawling. Data Knowl. Eng. **59**(2), 270–329 (2006)
18. Diligenti, M., Coetzee, F., Lawrence, S., Giles, C., Gori., M.: Focused Crawling using context graphs. In: Proceedings of 26th International Conference on Very Large Databases (VLDB 2000), pp. 527–534 (2000)
19. Chen, Y.: A novel hybrid focused Crawling algorithm to build domain-specific collections. Ph. D. thesis, Virginia Polytechnic Institute and State University (2007)
20. Zhang, X., Zhou, T., Yu, Z., Chen, D.: URL rule based focused Crawlers. In: IEEE International Conference on e-Business Engineering (2008)
21. Chakrabarti, S., van den Berg, M., Dom, B.: Focused Crawling: a new approach to topic-specific Web resource discovery. In: 8th International WWW Conference, May 1999
22. Varelas, G., Voutsakis, E., Raftopoulou, P., Petrakis, E.G.M., Milios, E.: Semantic similarity methods in WordNet and their application to information retrieval on the web. In: 7th ACM International Workshop on Web Information and Data Management (WIDM 2005), Bremen Germany (2005)
23. Liu, B.: Web data mining, from Chapter 6, 7, 8, pp. 183–235, 237–270, 273–318. Springer, Berlin (2007)
24. Bhatia, M.P.S., Gupta, D.: Discussion on web Crawlers of search engine. In: Proceedings of 2nd National Conference on Challenges and Opportunities in Information Technology (COIT-2008)
25. Soon, L.K., Ku, Y.E., Lee, S.H.: Web Crawler with URL signature—a performance study. In: 4th Conference on Data Mining and Optimization (DMO) (2012)
26. Kim, S.J., Jeong, H.S., Lee, S.H.: Reliable evaluations of URL normalization. In: Proceedings of the 2006 International Conference on Computational Science and its Applications (ICCSA), pp. 609–617 May 2006
27. Lee, S.H., Kim, S.J., Hong, S.H.: On URL normalization. In: Proceedings of the 2005 International Conference on Computational Science and its Applications (ICCSA), pp. 1076–1085, Singapore, May 2005
28. Berners-Lee, T., Fielding, R., Masinter, L.: Uniform resource identifier (URI): general syntax. Available at http://gbiv.com/protocols/uri/rfc/rfc3986.html
29. Garcia, E.: Vector models based on normalized frequencies. Mi Islita. Retrieved 17 Aug 2012 (2006)
30. Yongsheng, Y., Hui, W.: Implementation of focused Crawler, COMP 630D Course Project Report (2000)

Chapter 10
An Optimized Structure Filtered-x Least Mean Square Algorithm for Acoustic Noise Suppression in Wireless Networks

Asutosh Kar and Mahesh Chandra

Abstract In order to get distortionless signal, the noise that presents in original intended signal must be canceled. Various methods have been proposed to reduce acoustic noise which includes noise barriers, noise-absorbing circuits, and filters, whereas one of the effective ways of noise suppression is the use of adaptive filters by which continuous weight adaptation is done till the noise is minimized. Some of the basic noise reduction adaptive algorithms include least mean square (LMS) algorithm, normalized LMS algorithm, and recursive least square algorithm. The performance of the LMS algorithm is degraded in active noise control due to the presence of a transfer function in the auxiliary path following the adaptive filter. It degrades the rate of convergence and increases the residual power. To ensure convergence of the algorithm, the filtered-x least mean square (FX-LMS) algorithm finds its application where input to the error correlator is filtered by a copy of the auxiliary error path transfer function, whereas the FX-LMS results in very slow convergence performance in case of high tap length requirement for acoustic noise cancelation applications. In this paper, an improved pseudo-fractional tap length selection algorithm is proposed in context with the FX-LMS algorithm to find out the optimum structure of the acoustic noise canceler which best balances the complexity and steady-state performance.

Keywords Adaptive algorithm · Tap length · Active noise control · Least mean square · Mean square error · FX-LMS

A. Kar (✉)
Department of Electronics and Telecommunication Engineering, IIIT, Bhubaneswar, India
e-mail: asutosh@iiit-bh.ac.in

M. Chandra
Department of Electronics and Communication Engineering, BIT, Mesra, Ranchi, India
e-mail: shrotriya69@rediffmail.com

S. Patnaik et al. (eds.), *Recent Development in Wireless Sensor and Ad-hoc Networks*, Signals and Communication Technology, DOI 10.1007/978-81-322-2129-6_10

10.1 Introduction

Active noise control (ANC) is based on the principle of destructive superposition of acoustic waves [1]. It states that the net displacement of the medium through which two or more waves are traveling is the sum of their individual wave displacements. The net amplitude of the resulting wave depends on the relative phase of the two waves. When the positive antinodes of one wave align with negative antinodes of the other, the combined wave displacement will be zero. The same principle holds for more complicated waveforms and sound fields. Using ANC concept in noise cancelation, a noise signal is generated which is correlated with the noise signal but opposite in phase. By adding both the signals that are the signals corrupted by noise and the generated noise signal which is correlated and opposite phase to the actual noise signal, original signal can be made noise free. The generation of the signal is controlled by an adaptive algorithm which adaptively changes the weight of the filter used. ANC is mainly classified into two categories, i.e., "feed-forward ANC" and "feedback ANC." In feedback ANC method, a controller is used to modify the response of the system. The controller may be like an addition of artificial damping. But in feed-forward ANC method, a controller is used to adaptively calculate the signal that cancels the noise. In this paper, "feed-forward ANC" approach is implemented to cancel the noise. This is because the amount of noise reduction achieved by feed-forward ANC system is much more than that of feedback ANC system. The algorithm used to generate the required signal is filtered-X least mean square (FX-LMS) algorithm.

A single-channel feed-forward ANC system is comprised of a reference sensor, a control system, a canceling loudspeaker, and an error sensor. The primary noise present is measured by the reference sensor and is canceled out around the location of the error microphone by generating and combining an antiphase canceling noise that is correlated with the spectral content of the unwanted noise [2]. The reference sensor produces a reference signal that is "feed-forward" to the control system to generate "control signals" in order to drive the canceling loudspeaker to generate the canceling noise. The error microphone measures the residual noise after the unwanted noise and controls signal. It combines and sends an error signal back to the controller to adapt the control system in an attempt to further reduction of error. The control system adaptively modifies the control signal to minimize the residual error. The most famous adaptation algorithm for ANC systems is the FX-LMS algorithm [3] as mentioned earlier, which is a modified and improved version of the least mean square (LMS) algorithm.

However, FX-LMS algorithm is widely used due to its stability, but it has slower convergence. Various techniques have been proposed to improve the convergence of FX-LMS algorithm with some parameter trade-offs. Other versions of the FX-LMS include filtered-x normalized LMS (FX-NLMS) [4], leaky FX-LMS [5], and modified FX-LMS (MFX-LMS) [6]. However, the common problem with all of these algorithms is the slow convergence rate, especially when there is large number of weights. To overcome this problem, more complex algorithms such as filtered-x recursive least square (FX-RLS) [7] or filtered-x affine projection (FXAP) [8] are used. These algorithms have faster convergence rate compared to the FX-LMS; however, they involve matrix computations, and their real-time realizations might not be cost-effective. The reason for slow convergence in maximum cases of FX-LMS is heavy tap length requirement by the adaptive filter used in the identification framework. The total number of weights in an adaptive filter is represented as tap length. In many current designs where adaptive filters are employed, the tap length is made fixed at some compromise value which makes it less effective in dynamic time-varying scenarios. The perfect tap length matching best balances the complexity and steady-state performance of the adaptive filter in a system identification application which deals with nonlinearly varying unknown room impulse response. Hence, in this paper, an attempt has been made to find the optimized tap length of the adaptive filer which will increase the convergence performance without adding any extra structural or computation complexity to the overall design. An improved optimized structure FX-LMS algorithm is presented for acoustic noise cancelation in a time-varying environment. The algorithm analysis is made to find the variable parameters that best adjust the steady-state performance as well as the convergence speed in a time-varying environment under various noise conditions.

10.2 Problem Formulation

Here, selecting acoustic noise problems is not only based on high-frequency noise, but are also dominated by low-frequency noise. Various passive noise control techniques such as noise barriers and absorbers do not work efficiently in low-frequency environments. ANC works very efficiently in case of low-frequency noise, and it is also cost-effective than bulky, heavy barriers and absorbers. ANC can be used to successfully attenuate noise in ducts, automobiles, aircraft, and other applications.

The basic LMS algorithm fails to depict desirable results in ANC applications [5]. This is due to the assumption made that the output of the filter is the signal perceived at the error microphone, which is not the case in practice. The presence of the A/D, D/A converters, actuators, and antialiasing filter in the path from the output of the filter to the signal received at the error microphone causes significant changes in the output signal. This demands the need to incorporate the effects of this secondary path function in the algorithm [6–8]. But the convergence of the LMS algorithm depends on the phase response of the secondary path, exhibiting ever-increasing oscillatory behavior as the phase increases and finally going unstable at 90° [1, 5]. The solution to this problem was either to employ an inverse filter in the cancelation path or to introduce a filter in the reference path, which is ideally equal to the secondary path impulse response. The former technique is referred to as the "filtered-error" approach, while the later is now known as the "filtered-reference" method, more popularly known as the FX-LMS algorithm.

The FX-LMS solution is by far the most widely used due to its stable as well as predictable operation [6], whereas in normal ANC framework, there is a need of long tap length adaptive filter, where tap length implies the total number of filter coefficients/weights [4]. It decreases the performance of the noise cancelation design by increasing the overall structural complexity in a time-varying scenario and results in undermodeling and adaption noise in case of mismatch of filter length [9]. It also reduces the convergence speed of FX-LMS algorithm up to a great extent. Hence, there is a need of structure optimization of adaptive filter in an ANC application employing FX-LMS algorithm. The requirement is to construct a suitable dynamic tap length selection algorithm not only to minimize the structural and computational complexity of the filter design but also to improve the overall system identification performance.

Type-1 (Too short order filter) Too few filter coefficients result in undermodeling. Suppose there is a normal impulse response from an acoustic arrangement where the intension is to identify the long non-sparse system. A too short filter will result in degraded noise cancelation performance [9, 10] and demonstrates the problem of insufficient modeling.

Type-2 (Too large order filter) The obvious drawback of a too long filter is slow convergence. It is not suitable to have a too long filter as it increases the filter design complexity and introduces adaptation noise due to extra coefficients. Suppose there are two filters of different lengths to model acoustic noise cancelation arrangement where the too long filter converges slower than the filter which has same number of coefficients as the acoustic system to be identified. Due to the mismatch of tap length, the error spreads all over the filter and the adaptive filter itself introduces echo in this case [10].

10.3 The FX-LMS Algorithm

(1) Feed-forward ANC System:

The basic idea of feed-forward ANC is to generate a signal (secondary noise), which is equal to a disturbance signal (primary noise) in amplitude and frequency, but has opposite phase. Combination of these signals results in cancelation of the primary (unwanted) noise. This ANC technique is well known for its use in canceling unwanted sound and is presented in Fig. 10.1 [2].

(2) ANC System Description Using Adaptive Filter:

In this paper, instead of the genuine adaptive algorithm block, an improved optimized structure FX-LMS algorithm approach is employed to suppress noise. The most popular feed-forward control algorithm for acoustic noise cancelation is FX-LMS algorithm [7], which is the main area of focus in this paper. The adaptive

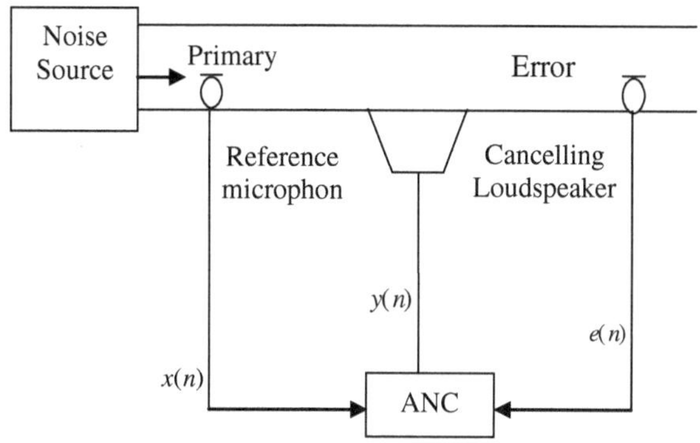

Fig. 10.1 Acoustic noise cancelation in a duct using feed-forward ANC system

filter action for any system identification application is shown in Fig. 10.2. The $x(n)$ represents the input signal. The filter weights are varied as per the error signal generated by the difference of output $y(n)$ from the desired signal $d(n)$. It monitors the adaptation process with a suitable adaptive algorithm.

The proposed optimized structure FX-LMS is presented in Fig. 10.3. A control signal is created by the FX-LMS algorithm by filtering the reference signal $x(n)$ with an adaptive control filter. The control filter is updated via a gradient descent search process until an ideal filter that minimizes the residual noise is found. This is

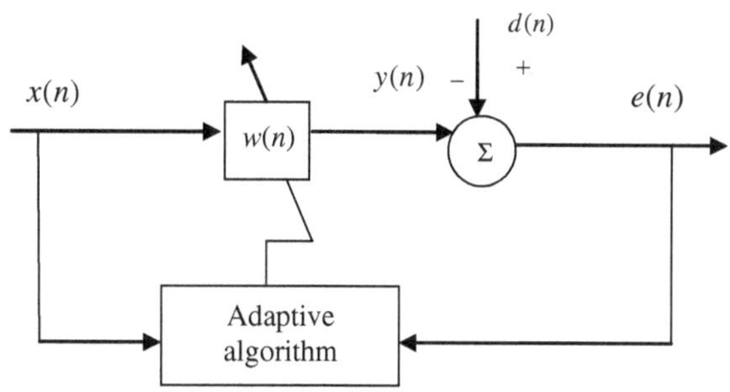

Fig. 10.2 System description using adaptive filter

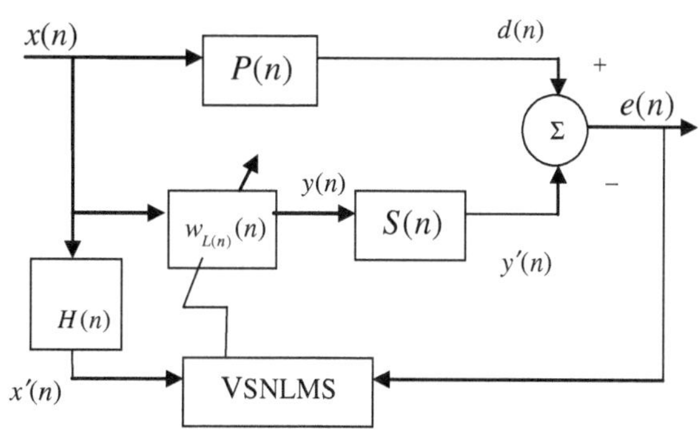

Fig. 10.3 Block diagram of optimized structure FX-LMS algorithm

because the existence of a filter in the auxiliary and the error path is shown, which generally degrades the performance of the LMS algorithm. Thus, the convergence rate is lowered, the residual power is increased, and the algorithm can even become unstable. In order to stabilize LMS algorithm, the reference signal $x(n)$ is filtered by an estimate of the secondary path transfer function $S(n)$ which is the propagation path from the controller to the error sensor giving the filtered-x signal. Hence, the algorithm is known as FX-LMS algorithm. The noise measured at the reference sensor propagates through the primary path represented by $P(n)$ and arrives at the error sensor or receiver as the signal, $d(n)$. This is the unwanted noise to be canceled, which is sometimes referred to as the "desired" signal, meaning the signal that the controller is trying to duplicate with opposite phase. The desired signal is to be correlated with the noise characterized by the reference. The LMS algorithm then generates control signal at the canceling loudspeaker by filtering the reference signal $x(n)$ with an adaptive FIR control filter, $w(n)$. The control signal, $y(n)$, is the convolution $x(n)$ of with $x(n)$. The control signal is filtered by the secondary path transfer function, $S(n)$, and arrives at the error sensor as the output signal, $y'(n)$.The output signal combines with the unwanted noise to give the residual error signal, $e(n)$, measured by the error sensor. An adaptive process searches for the optimal coefficients for the control filter which minimizes the residual error. This optimal filter is designated w_{opt}. In this proposed structure, the most important point to notice is the tap length optimization of the adaptive filter and the new weight update algorithm where all the parameters are made dynamic in order to support the tap length variation. If the noise environment is changed, then the filter length $L(n)$ automatically optimizes itself to a suitable value which avoids the occurrence of too short and too long filter as described in Sect. 10.2. Hence, the proposed structure works for narrow-band as well as wide-band noise, inside car to a big conference hall environment. Each time, the designer should not randomly fix any tap length for the filter because the filter itself adjusts its weight values as per the algorithm proposed in Sect. 10.4.

10.3.1 Implementation of FX-LMS Algorithm

Step: 1 For the ANC system in Fig. 10.3, containing a secondary path transfer function $S(n)$, the residual error can be expressed as follows:

$$e(n) = d(n) - y'(n) \tag{10.1}$$

where $y'(n)$ is the output of the secondary path $S(n)$.

Step: 2 If $S(n)$ is assumed as an IIR filter with denominator coefficients $[a_0 \ldots a_n]$ and numerator coefficients $[b_0 \ldots b_{M-1}]$, then the filter output $y'(n)$ can be written as the sum of the filter input $y(n)$ and the past filter output,

$$y'(n) = \sum_{i=1}^{N} a_i y'(n-i) + \sum_{j=0}^{M-1} b_j y(n-j) \tag{10.2}$$

Step: 3 It can be achieved in a similar way as done previously, so that the gradient estimate becomes

$$\nabla\hat{\xi}(n) = -2x'(n)e(n), \tag{10.3}$$

where

$$x'(n) = \sum_{i=1}^{N} a_i x'(n-i) + \sum_{j=0}^{M-1} b_j x(n-j) \tag{10.4}$$

In practical applications, $S(n)$is not exactly known; therefore, the parameters a_i and b_j are the parameters of the secondary path estimate $\hat{S}(n)$.

Step: 4 The weight update equation of the FX-LMS algorithm is as follows:

$$w_{L(n)}(n+k) = w_{L(n)}(n) + \mu_{L(n)}(n) \sum_{i=n}^{n+k-1} e_{L(n)}(i) X_{L(n)}(i) \tag{10.5}$$

Equation (10.5) is the filter's new weight update equation of the proposed optimized structure FX-LMS algorithm, which is described in the next section. As it can be seen that the equation is identical to the LMS algorithm but here instead of using $x(n)$, i.e., the reference signal, $x'(n)$, is used, which is the filtered output of the original signal, and it is called as "filtered-x signal" or "filtered-reference signal." Due to this change in the FX-LMS algorithm, the stability of FX-LMS algorithm is much more compared to the stability of LMS algorithm in ANC environment. The FX-LMS algorithm is very tolerant to modeling errors in the secondary path estimate. The algorithm converges when the phase error between $S(n)$ and $H(n)$ is smaller than ±90°. Phase errors less than 45° do not significantly affect performance of the algorithm [8]. The gain applied to the reference signal by filtering it with $S(n)$ does not affect the stability of the algorithm and is usually compensated by modifying the convergence step size.

Convergence is slowed down though, and the phase error increases. The reference signal, filtered by an estimate of the secondary path transfer function, gives the filtered-x signal that is used in the control filter update. $H(n)$ is estimated through a process called system identification. Band-limited white noise is played through the control speaker(s), and the output is measured at the error sensor. The measured impulse response is obtained as a finite impulse response (FIR) filter $S(n)$ in the time domain. The coefficients of $S(n)$ are stored and used to prefilter the reference signal and give the input signal to the LMS update.

The performance of the FX-LMS adaptation process is dependent on a number of factors. Those factors include the characteristics of the physical plant to be controlled, the secondary path impulse response, and the acoustic noise bandwidth. Hence, the fixed-length FX-LMS does not function properly if the secondary path has a long impulse response and/or the acoustic noise has a wide bandwidth [6]. Among the design parameters, the length of the transversal filter used as the ANC controller plays a great role. The steady state accomplishment of the adaptive filter is achieved with a higher tap-length. Whereas a larger order degrades the convergence of FX-LMS adaptation algorithm. So the filter length should be chosen carefully. It motivates research to find out the optimum length that best balances the system performance in a time-varying scenario. After the filter length is set, the maximum achievable performance is limited by a scalar parameter called the adaptation step size. The rules governing influences of step size on the LMS adaptation process can be used as rules of thumb for the adjustment of step size in the FX-LMS algorithm.

These rules can be stated as follows:

1. There is an upper bound for step size beyond which the process becomes unstable.
2. The convergence rate has a direct relationship with step size.
3. The steady-state performance has an indirect relationship with step size.

Based on these rules, appropriate step size can be chosen. As shown in the previous section, it is advantageous to choose a large value of μ because the convergence speed will increase. On the contrary, too large a value of μ will cause instability. It is experimentally determined that the maximum value of the step size that can be used in the FX-LMS algorithm is approximately [6]

$$\mu_{\max} = \frac{1}{p_{x'}(L + \Delta)} \tag{10.6}$$

where $p_{x'} = E[x'^2(n)]$ is the mean square value or power of the filtered-reference signal $x'(n)$, L is the number of weights, and Δ is the number of samples corresponding to the overall delay in the secondary path. This delay is determined by the phase error between the secondary path and secondary path estimate. But this phase error by definition is unknown, and hence, it is impossible to give an exact maximum value of μ. The step size in (10.6) does not result in a more converged FX-LMS algorithm as it uses a fixed tap length randomly fixed for a predefined problem. Hence, in this paper, as the tap length is a variable, hence, step size is made dependent on the variable tap length to result in good convergence as the filter order varies.

After convergence, the filter weights $w(n)$ vary randomly around the optimal solutionw_{opt}. This is caused by broadband disturbances, such as measurement noise and impulse noise on the error signal.

These disturbances cause a change of the estimated gradient $\nabla\hat{\xi}(n)$, because it is based only on the instantaneous error. This results in an average increase of the MSE, which is called the excess MSE and is defined as

$$\xi_{\text{excess}} = E[\hat{\xi}(n)] - \xi_{\min} \tag{10.7}$$

This excess MSE is directly proportional to step size. It can be concluded that there is a design trade-off between the convergence performance and the steady-state performance. A larger value of step size gives faster convergence but gives bigger excess MSE and vice versa. Another factor that influences the excess MSE is the number of weights, which is optimized in the next section.

10.4 Proposed Structure Optimization of FX-LMS

In this paper, the design problem is related to optimizing tap length-related criterion. The idea is to search the optimum filter length by establishing the cost function. If the difference of the MSE output of any two consecutive taps falls below a very small positive value, when the order is increased, then it can be concluded that adding extra taps does not reduce the MSE. Let define $D_L = Q_{L-1}(\infty) - Q_L(\infty)$ as the difference between the converged MSE when the filter order is increased from $L-1$ to L. Now, the optimum tap length can be defined as $\bar{L}$ that satisfies

$$D_L \leq \delta \quad \text{for all} \quad L \geq \bar{L} \tag{10.8}$$

where δ is a very small positive number set pertaining to the system requirement and $\min\{L|Q_{L-1} - Q_L \leq \delta\}$ is the cost function with respect to the filter order L. In many cases, pseudo-optimum filter order is observed.

An ANC model is considered in which both the optimum tap length $L_{\text{tpt}}(n)$ and coefficient of the unknown room impulse pertaining to that $w_{L_{\text{opt}}}(n)$ are to be identified. The weight update equation in the proposed algorithm is simplified by using segment-exact method that updates the filter coefficients every k sample with respect to variable tap length $L(n)$ [9, 11].

$$w_{L(n)}(n+k) = w_{L(n)}(n) + \mu_{L(n)}(n) \sum_{i=n}^{n+k-1} e_k^L(i) x_{L(n)}(i) \tag{10.9}$$

where n is the time index, μ' is a constant, and $L(n)$ is the instantaneous variable adaptive tap length obtained from the proposed fractional tap length estimation algorithm. $w_{L(n)}, x_{L(n)}$ are the weight and input vector pertaining to $L(n)$. $e_k^L(n)$ is the segmented error with respect to $L(n)$ defined as follows:

$$e_k^L(n) = d(n) - w^T_{L(n),1:k} x_{L(n),1:k}(n) \tag{10.10}$$

and

$$d(n) = w_{L_{opt}}^T(n)x_{L_{opt}}(n) + t(n) \tag{10.11}$$

where $1<k<L(n)$ and $w_{L(n),1:k}^T, x_{L(n),1:k}(n)$ are the weight and input vectors, respectively, consisting of first k coefficient of variable structure adaptive filter with $L(n)$ coefficients. Similarly, $w_{L_{opt}}^T(n), x_{L_{opt}}(n)$ are the weight and input vectors with tap length of L_{opt} and $t(n)$ is the system noise. The modified normalized LMS algorithm is sampled by employing tap length-varying step size,

$$\mu_{L(n)}(n) = \frac{\mu'}{\sigma_{X,n+k-1}^2[2 + L(n)]} \tag{10.12}$$

where $\sigma_{X,n+k-1}^2 = X_{L(n)}^T(n+k-1)X_{L(n)}(n+k-1)$ is the variance of input pertaining to $L(n)$ and $n + k - 1$ initial coefficients.

The mean square of the segmented error is represented as follows:

$$Q_k^{(L)}(n) = E[(e_k^L(n))^2] \tag{10.13}$$

Hence, the cost function for tap length optimization can be defined as the difference of MSE with an error spacing of $\Delta_L(n)$, [9]

$$\min\{L|Q_{L-\Delta_L(n)}^{(L)} - Q_L^{(L)}(n) \le \xi\} \tag{10.14}$$

where ξ is a small positive number whose value is determined by the system requirements according to the analysis in [9, 11].

However, different $\Delta_L(n)$ are needed for different applications, whereas for a certain application, it can be easily decided in advance according to the noise conditions.

Now, the algorithm for tap length adaptation in a time-varying environment is defined as follows:

$$L_{nf}(n+1) = [L_{nf}(n) - \Psi_n] + [(e_{k,L(n)}^L(n))^2 - (e_{k,L(n)-\Delta_L(n)}^L(n))^2]\bar{\Psi}_n \tag{10.15}$$

Finally, the tap length $L(n + 1)$ in the adaptation of filter weights for next iteration can be formulated as follows:

$$\begin{array}{ll} L(n+1) = \langle L_{nf}(n) \rangle & \text{if} |L(n) - L_{nf}(n)| > \frac{\Psi_n}{\bar{\Psi}_n} \\ L(n) & \text{otherwise} \end{array} \tag{10.16}$$

$L_{nf}(n)$, the tap length, can take fractional values. As the actual tap length of the adaptive filter cannot be a fractional value, so $L_{nf}(n)$ is rounded to the nearest integer value to get the optimum tap length. In (10.15), the factor Ψ_n is the leakage factor which prevents the order to be increased to an unexpectedly large value and

Ψ_n is the step size for filter order adaptation. In [11], the value of $(\Psi_n, \bar{\Psi}_n)$ was based on setting a random leaky factor which performed well for FIR systems, especially for the issues of acoustic echo cancelation. In this paper, a unique method for setting these parameters has been defined which can be applicable both for infinite impulse response and for FIR systems [12].

$$\Psi_n = \min(\Psi_{n,\max}, \Psi_n(i+1)) \tag{10.17}$$

$$\Psi_n(i+1) = \frac{\tilde{e}^2_{k,L(n)}(i+1)}{\tilde{e}^2_{k,L(n)}(i+1) + \Delta_{L_{ss}}(i)} \tag{10.18}$$

where

$$\tilde{e}^L_{k,L(n)}(i+1) = f\tilde{e}^L_{k,L(n)}(i) + (1-f)\tilde{e}^L_{k,L_{\text{opt}}(n)}(i+1) \tag{10.19}$$

$\Delta_{L_{ss}}$ defines the variable error spacing parameter at steady-state tap length $L_{ss}(\infty)$and f is a partial weight factor. At the steady state, [11–13]

$$\Psi_n \rightarrow \frac{(1-f)\sigma_t^2}{(1+f)\Delta_{L_{ss}}(\infty)} \tag{10.20}$$

Similarly, the adaptation step size depends on the bias between MSE values with a Δ_L difference. If the difference is more, then adaptation should be slow and vice versa.

$$\Psi_n = \min(\bar{\Psi}_{n,\max}, \tau\bar{\Psi}_{n,\max}) \quad 0<\tau\leq 1 \tag{10.21}$$

where $(\Psi_{n,\min}, \Psi_{n,\max})$ and $(\bar{\Psi}_{n,\min}, \bar{\Psi}_{n,\max})$ can be fixed as [12].

The combined tap length adaptation for both the defined occasions can be specified by employing a direction vector $\phi = \text{sgn}(X_n)$ where $X_n = L(n) - L(n-1)$. Now, (10.15) is modified as

$$L_{nf}(n+1) = [L_{nf}(n) - \Psi_n] + \phi[Q^{(L)}_{L-\Delta_L}(n) - Q^{(L)}_L(n)]\bar{\Psi}_n \tag{10.22}$$

Comparison of available structures and adaptation algorithms used in ANC frame is shown in Table 10.1. From the comparison table, it can be easily shown that however, FX-LMS algorithm has slow convergence when applied to noise cancelation, but due to its simple time realization and low computational complexity, it is widely used. On the other hand, FX-RLS algorithm has fast convergence, but it involves huge computational complexity and it also has difficult real-time realization. So FX-RLS algorithm is not a preferred algorithm to update weight vector in case of ANC scenario. Both frequency-domain FX-LMS and sub-band FX-LMS algorithm are only applicable to narrow-band noise. Hence, considering all aspects, the proposed optimized structure FX-LMS algorithm performs better in ANC environment than other three algorithms mentioned.

Table 10.1 Comparison of different variants of FX-LMS for ANC

Algorithms	Strengths	Weaknesses
FX-LMS	1. Simple real-time realization 2. Low computational complexity	1. Slow convergence
FX-RLS	1. Fast convergence	1. Huge computational complexity
	2. Low steady-state residual noise	2. Difficult real-time realization
Frequency-domain FX-LMS	1. Simple real-time realization 2. Low computational complexity 3. Fast convergence	1. Only suitable for narrow-band noise
Sub-band FX-LMS	1. Low computational complexity 2. Fast convergence	1. Only suitable for narrow-band noise
Proposed optimized structure FX-LMS	1. Fast convergence 2. Very less structural complexity 3. Low steady-state residual noise 4. Suitable for wide variety of noise signals	1. Difficult real-time realization in highly dynamic noisy environment

10.5 Simulation Setup and Results

MATLAB platform is chosen for simulation purpose. The room impulse response is measured for an enclosure of dimension $10 \times 10 \times 8$ ft^3 using an ordinary loudspeaker and an unidynamic microphone kept at a distance of 1 ft which has been sampled at 8 kHz. The channel response is obtained by applying a stationary Gaussian stochastic signal with zero mean and unit variance as input on the measured impulse response. The primary and secondary path impulse response with respect to time is presented in Figs. 10.4 and 10.5, respectively. The secondary propagation path is the path that antinoise takes from the output loudspeaker to the error microphone within the quiet zone. For this ANC task, a sampling frequency of 8,000 Hz should be used. The first task in ANC is to estimate the impulse response of the secondary propagation path. This is usually performed prior to noise control using a random signal played through the output loudspeaker, while the unwanted noise is not present. The actual signal along with the estimate in FX-LMS and the resulting error is represented in Fig. 10.6. Typically, the length of the secondary path filter estimate is not as long as the actual secondary path and need not be for adequate control in most cases. But finding the exact length which best adjusts the system performance is a trivial task. Hence, the tap length optimization algorithm can be applied to find the dynamic filter length that best adjusts the system performance in a time-varying scenario.

Considering Fig. 10.3, the noise $x(n)$ is propagating from the source to the sensor, through the plant $P(n)$.

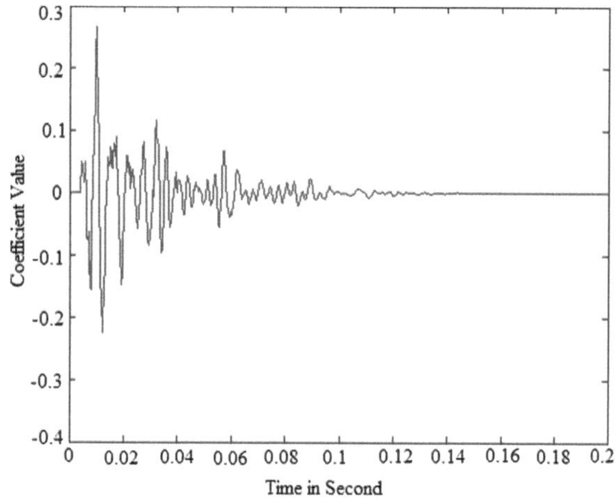

Fig. 10.4 Primary path impulse response

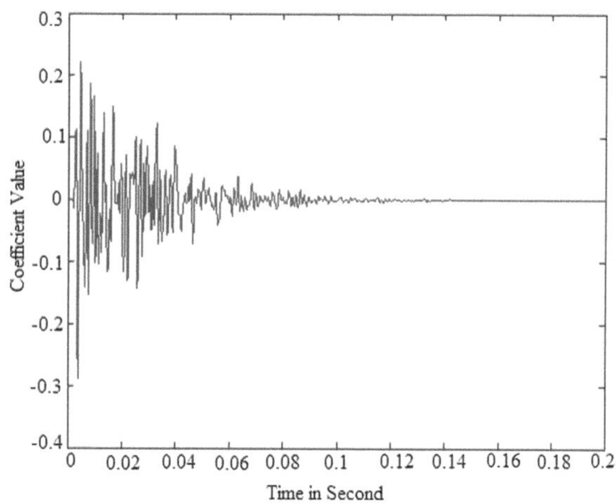

Fig. 10.5 Secondary path impulse response

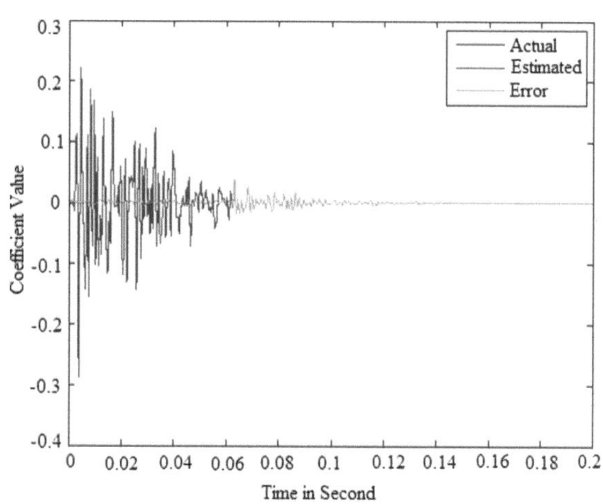

Fig. 10.6 Estimated and error signal for the actual signal

The sensor measures the arriving noise as $d(n)$. To reduce noise, another "noise" $y(n)$ is generated using the controller $w(n)$. It is hoped that the generated signal destructively interferes $x(n)$. This means that the controller has to be a model of the propagation medium $P(n)$. The variable structure FX-LMS algorithm is applied to adjust the controller coefficient. However, there is also medium $S(n)$ that stays between the actuator and sensor. So, to make the solution right, the adjustment process should be compensated using $H(n)$, which is an estimate of $S(n)$.

The compensation of medium $S(n)$ using $H(n)$ is shown in Fig. 10.7. Then, Fig. 10.8 shows the clean speaker signal, signal corrupted by car noise signal, and the filtered signal.

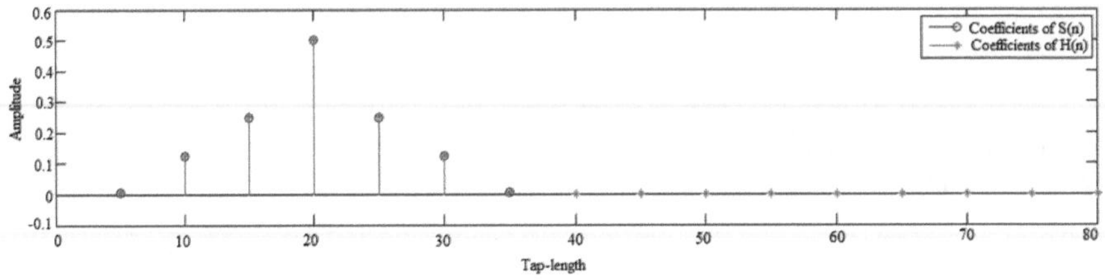

Fig. 10.7 Amplitude variation with respect to tap length for $S(n)$ and $H(n)$

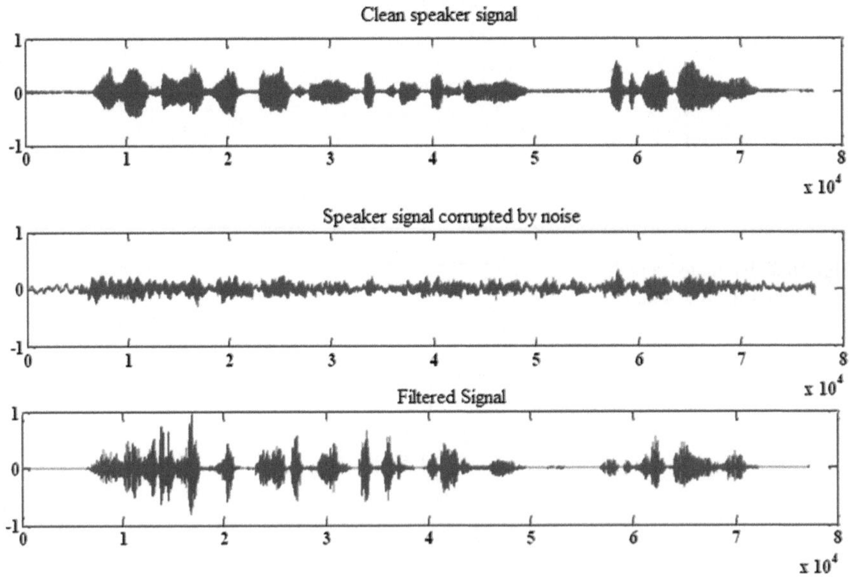

Fig. 10.8 The noise cancelation process from the actual signal corrupted by car noise

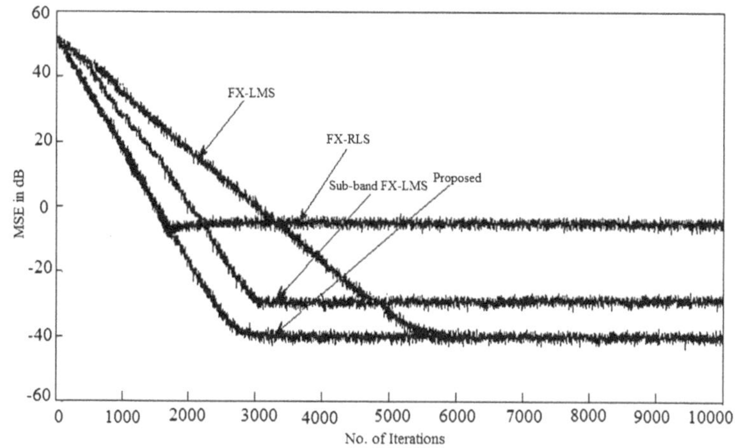

Fig. 10.9 MSE versus number of iterations (SNR = 35 dB)

The proposed variable structure FX-LMS algorithm best adjusts the system performance in comparison with the fixed-length FX-LMS, FX-RLS, and sub-band FX-LMS algorithm as shown in Fig. 10.9 in the MSE variation with respect to iterations. It can be observed that the FX-RLS shows better convergence speed over all the other variants of FX-LMS algorithms, but it lacks in MSE performance. Similarly, the sub-band FX-LMS algorithm shows slightly better MSE behavior compared to fixed-length FX-LMS but fails to achieve the performance depicted by the proposed dynamic structure FX-LMS algorithm. The proposed algorithm out performs all the algorithms if the tracking performance is considered for an enclosure dislocation as depicted in Fig. 10.10.

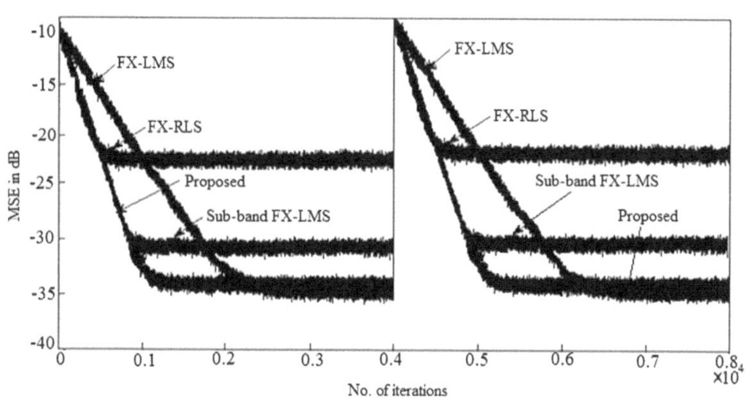

Fig. 10.10 Tracking performance comparison (SNR = 20 dB)

10.6 Conclusion

FX-LMS algorithm is widely used in acoustic noise cancelation environment due to its simple real-time implementation and low computational complexity. However, the algorithm has slow convergence, but the optimization in tap length in the existing algorithm is done to improve the convergence. The variable structure FX-LMS algorithm has greater stability than other algorithms used in case of acoustic noise cancelation as shown in the paper. The design may further be studied for applications such as channel equalization and echo reduction where typically large tap length filters are utilized. High-noise and double-talk detection issues can further be analyzed for more improved analysis.

References

1. Elliot, S.J., Nelson, P.A.: Active noise control. IEEE Signal Process. Mag. **10**, 12–35 (1993)
2. Kuo, Sen M., Morgan, Dennis R.: Active noise control: a tutorial review. Proc. IEEE **87**, 943–973 (1999)
3. Rupp, M., Sayed, A.H.: Robust FxLMS algorithms with improved convergence performance. IEEE Trans. Speech Audio Process. **6**(1), 78–85 (1998)
4. Warnaka, G.E., Poole, L.A., Tichy, J.: Active acoustic attenuators. US Patent 4473906, 1984
5. Elliott, S., Stothers, I.M., Neslson, P.A.: A multiple error LMS algorithm and its applications to active control of sound and vibration. IEEE Trans. Acoust. Speech Signal Process. **35**, 1423–1434 (1987)
6. Kar, A., Chand, A.P., Chandra, M.: An improved filtered-x least mean square algorithm for acoustic noise suppression. In: Advanced Computing, Networking and Informatics, vol. 1, pp. 25–32. Springer-Verlag, Germany (2014)
7. Oppenheim, A., Weinstein, E., Zangi, K., Feder, M., Gauger, D.: Single-sensor active noise cancellation. IEEE Trans. Speech Audio Process. **2**(2), 285–290 (1994)
8. Gonzalez, A., Ferrer, M., de Diego, M., Pinero, G.: Fast filtered-x affine projection algorithm for active noise control. IEEE Workshop on Applications of Signal Processing to Audio and Acoustics, pp. 162–165 (2005)
9. Kar, A., Chandra, M.: A minimized complexity dynamic structure adaptive filter design for improved steady state performance analysis. Int. J. Comp. Vis. Robot. **3**(4), 326–340 (2013)
10. Kar, A., Chandra, M.: Pseudo-fractional tap-length learning based applied soft computing for structure adaptation of LMS in high noise environment. In: Soft Computing Techniques in Engineering Applications, pp. 115–129. Springer-Verlag, Germany (2014)
11. Kar, A., R.Nath, Alaka Barik.: A VLMS based pseudo-fractional order estimation algorithm. In: Proceedings of ACM Sponsored International Conference on Communication, Computing and Security, pp. 119–123 (2011)
12. Gang, Y., Li, N., Chambers, J.A.: Steady-state performance analysis of a variable tap-length LMS algorithm. IEEE Trans. Signal Process. **56**, 839–845 (2008)
13. Mayyas, K.: Performance analysis of the deficient length LMS adaptive algorithm. IEEE Trans. Signal Process. **53**(8), 2727–2734 (2005)
14. Kar, A., Chandra, M.: A novel variable tap-length learning algorithm for low complexity, fast converging stereophonic acoustic echo cancellation. Int. J. Inf. Commun. Technol. **6**(3/4), 309–325 (2014)

Chapter 11
An Exhaustive Comparison of ODMRP and ADMR Protocols for Ad hoc Multicasting

Ajit Kumar Nayak, Srikanta Patnaik and Rajib Mall

Abstract Currently, multicasting applications such as gaming and chatting are gaining popularity in wireless ad hoc networks. In this work, a rigorous comparison of two existing major protocols ODMRP and ADMR has been made from many different aspects. Also, the performance of each protocol is carefully examined and the conclusion is drawn accordingly. We found that ODMRP is a better protocol as compared to ADMRP in many different scenarios.

Keywords Ad hoc multicast · ODMRP · ADMR · Network performance

11.1 Introduction

Multicasting is the transmission of messages simultaneously to a number of destinations that are identified by one single group address. Efficient routing for multicasting is much more difficult than unicasting as the effective group management needs to be addressed. Multicasting is needed for many popular applications that are characterized by a close degree of collaborations among hosts such as group messaging, gaming, conferencing, and chatting. Such applications are especially

A.K. Nayak (✉)
Department of Computer Science and Information Technology, Institute of Technical Education and Research, Siksha 'O' Anusandhan University, Khandagiri Square, Bhubaneswar, Odisha, India
e-mail: ajitnayak@soauniversity.ac.in

S. Patnaik
Department of Computer Science and Engineering, Institute of Technical Education and Research, Siksha 'O' Anusandhan University, Khandagiri Square, Bhubaneswar, Odisha, India
e-mail: patnaik_srikanta@yahoo.co.in

R. Mall
Department of Computer Science and Engineering, Indian Institute of Technology Kharagpur, Kharagpur 721302, West Bengal, India
e-mail: rajib@cse.iitkgp.ernet.in

S. Patnaik et al. (eds.), *Recent Development in Wireless Sensor and Ad-hoc Networks*, Signals and Communication Technology, DOI 10.1007/978-81-322-2129-6_11

important for users when they are mobile and a community of similar interest is formed on demand using handheld devices. For example, a person waiting for a train may need to utilize his/her time by remotely playing a game or discussing about current issues with others with the help of services available on his/her handheld device.

In recent years, a number of novel multicast protocols of different styles have been proposed for ad hoc networks. In this chapter, we present our results concerning comparison of two well-known protocols for wireless multicasting. One of the considered protocols is "on-demand multicast routing protocol" (ODMRP) [1], and the other one is "adaptive demand-driven multicast routing protocol" (ADMR) [2]. ODMRP is a mesh-based protocol, whereas ADMR uses a tree-based technology for routing. A few works have been reported in the literature to compare the performance of ad hoc multicast protocols [3–6]. Our objective is to investigate comparative results on these two protocols in more detail to analyze their relative advantages that would help us to synthesize a new protocol such that it would have the benefits of both. These two protocols are exhaustively simulated to investigate their performances under various execution environments. The important parameters were measured by varying characteristics such as mobility speed, number of multicast senders, multicast group size, network traffic load, and different pause times. We believe that this study can help to decide a routing scheme under a particular working environment.

The rest of the chapter is organized as Sect. 11.2 provides an overview of some recent multicast protocols in detail. Section 11.3 presents the general overview and working principles of the two most recent ad hoc multicast protocols ODMRP and ADMR. Section 11.4 describes our simulation model such as radio model, mobility scenarios, communication scenarios, and performance metrics. Section 11.5 presents the results obtained from simulation experiments and the analysis of the results. Section 11.6 compares protocol performance, and finally, the summary is presented in Sect. 11.7

11.2 A Review of Multicast Protocols

Over the last two decades, numerous works have targeted providing efficient and robust multicast routing mechanisms for mobile ad hoc networks. But still, research in this area is far from complete. Most of the early works in ad hoc network routing were designed for *unicasting*. Multicasting in ad hoc networks was addressed over the last few years by using a variety of mechanisms [1, 7–28]. All the general-purpose multicast routing protocols presented so far may be classified into four categories based on how routes are created to the members of the group namely (i) tree-based approaches, (ii) mesh-based approaches, (iii) stateless multicast, and (iv) hybrid approaches.

Tree-based multicast is a very well-established concept in wired networks. But due to unpredictably varying network conditions in a MANET, the same protocols would not work efficiently. Different researchers have tried to extend the tree-based approach to provide efficient multicast routing in a MANET environment. In this approach, a tree structure is formed rooted at some specific node called the core or source or group leader. A prominent example is ad hoc multicast routing protocol utilizing increasing id numbers (AMRIS) [14]. It is an on-demand protocol that constructs a shared multicast delivery tree to support multiple senders and receivers. MAODV [21] is an extension to unicast routing protocol AODV and discovers multicast routes on demand using a broadcast route discovery mechanism. Lightweight adaptive multicast (LAM) [29] is another protocol designed from core-based tree [30] and ad hoc unicasting protocol TORA. Adaptive demand-driven routing protocol (ADMR) [2] attempts to reduce any non-on-demand components as much as possible from within the protocol. Multicast routing state is dynamically established and maintained only for active groups and only in nodes located between multicast senders and receivers.

Mesh-based multicast protocols may have multiple paths between source and destination pair. These protocols are very robust and best suited in an environment where the topology changes frequently. Routing of packets is done via alternate paths in case of link failure due to high mobility. The drawback is the requirement of excessive control messaging to maintain redundant path in the mesh and thus less efficient. The core-assisted mesh protocol (CAMP) [13] supports multicasting by creating a shared mesh for each multicast group. It assumes a mapping service, which provides routers with the address of groups identified by their names. On-demand multicast routing protocol (ODMRP) [1] is the most promising protocol in this segment. It is a mesh-based on-demand protocol that uses a forwarding group concept (only a subset of nodes forwards the multicast packets). A soft-state approach is taken in ODMRP to maintain multicast group members. No explicit control message is required to leave the group. In ODMRP, group membership and multicast routes are established and updated by the source on demand. Forwarding group multicast protocol (FGMP) [116] uses flooding with limited scope.

Stateless multicast schemes avoid the overhead of creating and maintaining the delivery tree/mesh with time. In stateless multicast, a source explicitly mentions the list of destinations in the packet header. Stateless multicast focuses on small group multicast and assumes the underlying unicast routing protocol to take care of forwarding the packet to respective destinations based on the addresses contained in the header. The main problem with this kind of routing is redundant transmission of the same packet to each destination separately. The differential destination multicast (DDM) protocol [27] is meant for small-multicast groups operating in dynamic networks of any size. Unlike other MANET routing protocols, DDM lets the source control multicast group membership. The source encodes multicast receiver addresses in multicast data packets using a special DDM data header. This variable-length destination list is placed in the packet headers, resulting in packets being self-routed toward their destinations using the underlying unicast routing protocol. It eliminates maintaining per-session multicast forwarding states at intermediate nodes and thus is easily scalable with respect to the number of sessions.

Hybrid approaches are designed to achieve better performance by combining the advantages of both tree- and mesh-based approaches. The tree-based approaches provide high data forwarding efficiency at the expense of low robustness, whereas mesh-based approaches provide better robustness (link failure may not trigger a reconfiguration) at the expense of higher forwarding overhead and increased network load. The ad hoc multicast routing protocol (AMRoute) [31] creates a bidirectional, shared tree using only group senders and receivers as tree nodes for data distribution. The protocol has two main components: mesh creation and tree setup. The mesh creation identifies and designates certain nodes as logical cores, and these are responsible for initiating the signaling operation and maintaining the multicast tree to the rest of the group members. A non-core node only responds to messages from the core nodes and serves as a passive agent. The selection of a logical core in AMRoute is dynamic and can migrate to any other member node depending on network dynamics and group membership.

Besides these, in [32], authors have tried to develop multicasting techniques to support both wired and wireless networks.

11.3 Protocol Overview

In this section, a brief overview of the two ad hoc multicast protocols is presented, that is, chosen for comparison.

11.3.1 On-Demand Multicast Routing Protocol (ODMRP)

ODMRP [1] creates a mesh of nodes (the "forwarding group") which forward multicast packets via flooding (within the mesh), thus providing path redundancy. ODMRP is an on-demand protocol; thus, it does not maintain route information permanently. It uses a soft-state approach in group maintenance. Member nodes are refreshed and needed and do not send explicit leave messages.

In ODMRP, group membership and multicast routes are established and updated by the source on demand. Similar to on-demand unicast routing protocols, a request phase and a reply phase comprise the protocol. When multicast sources have data to send, but do not have routing or membership information, they flood a JOIN QUERY packet. When a node receives a non-duplicate JOIN QUERY, it stores the upstream node ID (i.e., backward learning) and rebroadcasts the packet. When the JOIN DATA packet reached a multicast receiver, the receiver creates a JOIN REPLY and broadcasts to the neighbors.

As shown in Fig. 11.1, source "S" floods a JOIN QUERY to entire network to refresh membership. Receiving node "R" stores the backward learning into routing table and rebroadcasts the packet. Finally, when query reaches, a receiver creates a JOIN REPLY and broadcasts it to its neighbors.

When a node receives a JOIN REPLY, it checks whether the next node ID of one of the entries matches its own ID. If it does, the node realizes that it is on the

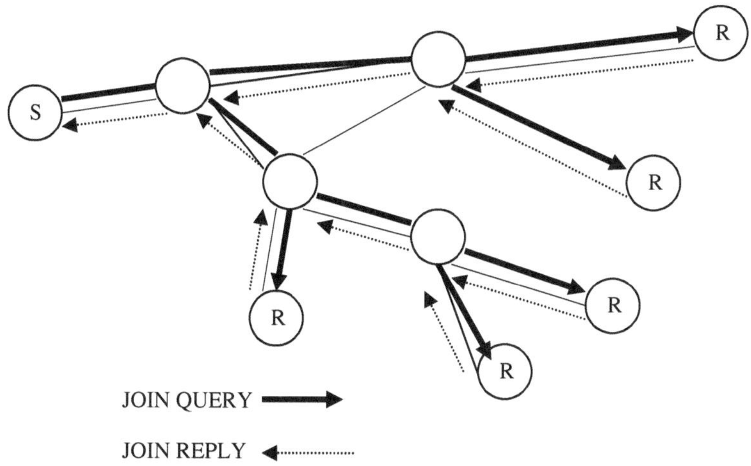

Fig. 11.1 Forwarding mesh in ODMRP

path to the source and thus is part of the forwarding group. Routes from sources to receivers build a mesh of nodes called "forwarding group." The intermediate node then broadcasts its own JOIN REPLY built upon matched entries. The JOIN REPLY is thus propagated by each forwarding group member until it reaches the multicast source via the shortest path. This process constructs (or updates) the routes from sources to receivers and builds a mesh of nodes, the forwarding group. Multicast senders refresh the membership information and update the routes by sending JOIN QUERY periodically.

In networks where global positioning system (GPS) [29] is available, ODMRP can be made adaptive to node movements by utilizing mobility prediction. By using location and mobility information supported by GPS, route expiration time can be estimate and receivers can select the path that will remain valid for the longest time. With the mobility prediction method, sources can reconstruct. Routes in anticipation of route breaks. This way, the protocol becomes more resilient to mobility. The price is, of course, the cost and additional weight of GPS.

The data transfer phase is identical for both versions. Nodes forward data if they are the forwarding nodes and the packet they receive is not a duplicate. Since all forwarding nodes relay data, redundant paths (when they exist) can help deliver data when the primary path becomes disconnected because of mobility.

Another unique property of ODMRP is its unicast capability. Not only can ODMRP coexist with any unicast routing protocol, it can also operate very efficiently as unicast routing protocol. Thus, a network equipped with ODMRP does not require a separate unicast protocol.

11.3.2 Adaptive Demand-Driven Multicast Routing Protocol (ADMR)

ADMR [2] is an on-demand tree-based multicast protocol. It neither uses network-wide control packet floods nor does it require periodic exchanges of messages for routing table update or neighbor sensing. Also, it does not rely on lower layers within the protocol stack to perform such functions; it performs both its route discovery and route maintenance functions on demand, and automatically prunes unneeded multicast forwarding states, and expires its multicast mesh when it detects that the multicast application has become inactive. When there are no multicast sources or receivers for a given multicast group G, ADMR does not generate any packet transmissions. If multicast receivers and sources for G exist, ADMR creates a source-based forwarding tree rooted at multicast sender S as in Fig. 11.2 and the multicast receivers for the group form the branches/leaves of the tree.

Packets with multicast destination containing ADMR header uses tree flooding. Source-specific forwarding enables the protocol to support source-specific multicast joins and to route along shorter paths than protocols that use group-shared forwarding. Packet forwarding along the ADMR source tree does not follow any predetermined sequence of hops, but instead each non-duplicate data packet is forwarded by each tree node to the multicast receivers. This type of forwarding increases robustness against packet loss due to collisions or broken links.

The multicast sources and receivers in ADMR cooperate to create the multicast source mesh. Each source floods its first data packet for a group, and each receiver responds to that flood with a RECEIVER JOIN packet which sets up forwarding state along the shortest path back toward the source. A flood-response cycle is initiated by each receiver when it first joins the group. To resolve partitions, multicast sources may occasionally flood a data packet, e.g., after every several tens of seconds. To detect broken links within the mesh, the ADMR routing layer at a multicast source monitors the traffic pattern of the multicast source application and includes the expected inter-arrival time.

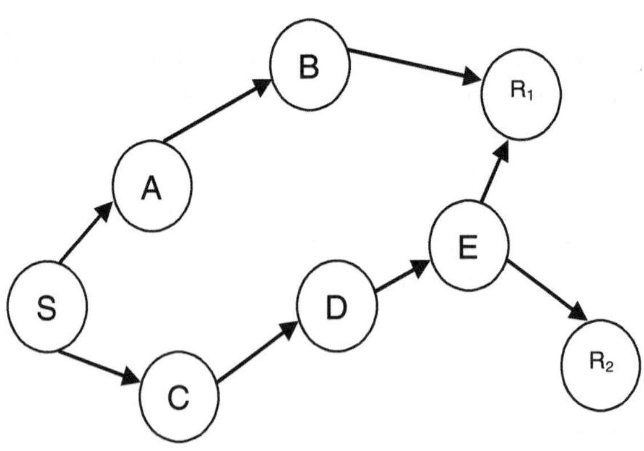

Fig. 11.2 Multicast tree in ADMR

Table 11.1 Summary of key features of ODMRP and ADMR

Protocols	ODMRP	ADMR
Configuration	Mesh	Tree
Loop free	Yes	Yes
Dependency on unicast	No	No
Periodic messaging	Yes	No
Control packet overhead	Yes	No

11.3.3 Protocols Summary

Table 11.1 summarizes key characteristics and properties of the protocols simulated. Note that ODMRP requires periodic messaging, whereas ADMR does not require.

11.4 Simulation Model

For the simulation, NS-2 [33] network simulator has been used. It is the discrete event packet-level simulator targeted at networking research along with the multicast extensions developed by CMU [34]. These extensions include implementations of models of signal strength, radio propagation, wireless medium contention, capture effect, and node mobility. We have used a free-space propagation model, in which the power of a signal attenuation is inversely proportional to the square of the distance between radios. The radio model is based on the 802.11 [35], which provides a 2-Mbps transmission rate and a nominal transmission range of 250 m. The MAC uses distributed coordinated function (DCF) of 802.11. DCF allows nodes to share the wireless medium using CSMA/CA.

The values of the main parameters of each protocol are listed in Tables 11.2 and 11.3, and have been set as reported in the corresponding literatures and/or suggested

Table 11.2 Simulation parameters for ADMR

Parameter	Value
MAX_NUM_RCV_JOIN	3
DEFAULT_EXPECTED_PKT_INTERVAL_TIME	0.2
DEFAULT_KEEPALLIVE_COUNT	10
DEFAULT_MULTIPLICATION_FACTOR	1
BEGIN_LOCAL_REPAIR_DELAY	0.2
LOCAL_REPAIR_TTL	2
NUM_MISSING_PKTS_TRIGGER_DISCUSSION	3
NUM_MISSING_PKTS_TRIGGER_EXPIRATION	15
TEMPORARY_STATE_EXPIRATION_TIMER	10

Table 11.3 Parameters for ODMRP

Parameter	Value
REFRESH_INTERVAL	3 S
JOIN_AGGREGATION_TIMEOUT	0.025 S
RREQ_RETRIES	2
FORWARDING_STATE_TIMEOUT	3x REFRESH_TIMEOUT
JOIN_REPLY_PASSIVE_ACK_TIME_OUT	0.035 S
MAX_NUM_JOIN_REPLY_RETRIES	3

by their designers. Some parameters are assumed in case of unavailability of information such as STATE_SETUP_TIMER and PASSIVE_ACK_TIMER for ADMR and maximum number of RREP retransmissions for ODMRP.

11.4.1 Mobility Scenarios

The experiments include networks with 100 nodes placed on a site with dimensions 2,000 × 2,000 m. Nodes in each scenario move according to the random waypoint model in which each node independently picks a random destination and speed from an interval (0, *Max Speed*) and moves toward the chosen destination at the selected speed. When the node reaches the destination, it stops for *pause time* seconds and then repeats the process. The pause time 0 is considered as continuous movement of nodes. The maximum speed is 20 m/s. Each simulation is run for 900 s. Multiple randomly generated scenarios with different seeds are conducted for each parameter combination, and each point in the graphs is the average of the results of these scenarios.

11.4.2 Communication Scenarios

The multicast sources begin to send data, and the multicast receivers join a multicast group at uniformly randomly chosen times between 0 and 180 s from the beginning of the simulation and remain in the multicast session until the end of the simulation. In all scenarios, constant bit rate (CBR) traffic generators send 512-byte packets. This packet size was chosen in order to increase the likelihood of congestion. The packet rates are chosen to continuously probe the routing ability of the protocols rather than to represent any particular application.

We have implemented many scenarios to properly investigate the performance of considered protocols under different network environments. We have varied the parameters such as mobility speed, number of multicast senders, multicast group size, network traffic load, pause time, and number of groups.

11.4.3 Performance Metrics

Protocol performance is evaluated using the following metrics, which are computed over the whole duration of the simulation:

- **Packet Delivery Ratio**: The ratio of packets sent by the multicast application that are received by the multicast receivers. For example, if there are 2 sources and 5 receivers, and each source sends 10 packets, and each receiver receives 12 multicast packets total, the packet delivery ratio would be (5 × 12)/(5 × 20), which is 0.6., i.e., packets actually received/packets supposed to be received.
- **Number of data packets transmitted per data packets delivered**: Data packets transmitted is the count of every individual transmission of data by each node over the entire network. This count includes the packets forwarded by intermediate nodes. In multicast, as a single transmission can deliver to multiple destinations, the measure may be less than one.
- **Control bytes transmitted per data byte delivered**: The total number of control bytes transmitted per number of data bytes delivered. This metric measures the efficiency of control message utilization toward data delivery.
- **Normalized Packet Overhead**: The number of control and data transmissions performed by the protocol per successfully delivered data packet. This metric evaluates the overall effort that the protocol expends for the delivery of each data packet. For example, multicast efficiency of 5 means that the protocol makes 5 packet transmissions on average for each data packet that is delivered to a multicast receiver.
- **Forwarding Efficiency**: The average number of times each originated data packet transmitted by the routing protocol. This metric represents the efficiency of the basic multicast forwarding within the routing protocol
- **Delivery Latency**: The average time between the transmission of a data packet from a multicast source and the time of its reception by a multicast receiver, averaged over all receivers.

11.5 Results and Discussions

All the simulation results are presented in the form of line graphs using Fig. 11.3 through Fig. 11.8. Each of these cases is discussed individually in following subsections.

11.5.1 Varying Mobility Speed

In this set of experiments, nodes do move in different speeds keeping all other features same for all simulations. Each node moves constantly with predetermined

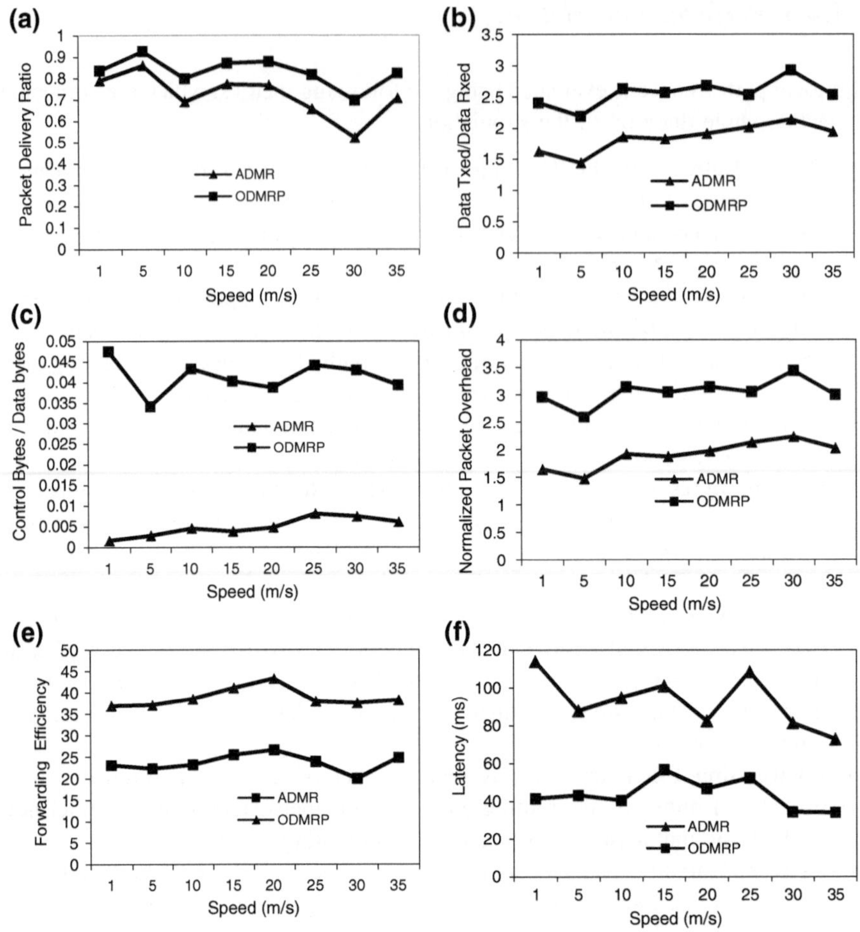

Fig. 11.3 Performance of protocols with changing speed

speed in random directions. The node movement speed was varied from 1 to 35 m/s or 126 km/h. The simulation was conducted for one multicast group, one source and 20 receivers per group. Source transmits 4 CBR packets/s with no pause time, i.e., nodes move without rest.

The packet delivery ratio has been compared for different speeds in Fig. 11.3a. As can be observed, ODMRP shows slightly better performance as compared to ADMR even in highly dynamic situations. The reason can be that, ODMRP provides redundant routes with a mesh topology and the chances of packet delivery to destinations remain high even when the primary routes are unavailable. The path redundancy enables ODMRP to offer only minimal data loss and to be robust to mobility, whereas the performance of ADMR is not so poor. The delivery ratio in case of ADMR is slightly less than ODMRP.

Number of data transmissions per data delivery to destinations: ODMRP has the more number of transmissions as shown in Fig. 11.3b because of redundant paths.

It can be observed that protocols using meshes ODMRP transmit more data packets than ADMR, which uses a forwarding tree. In fact, ODMRP transmits nearly as much data as flooding because it exploits multiple redundant routes for data delivery.

Control overhead: As given in Fig. 11.3c, the control bytes transmitted as compared to data bytes delivered is much more in case of ODMRP as compared to ADMR because of periodic updates. Also, in ODMRP, the control overhead remains relatively constant because no updates are triggered by mobility. JOIN DATA refresh interval was set constant to three seconds, and hence, no additional overhead is required as mobility increases. ADMR contributes less to control bytes as it does not require periodic updates.

Total number of packets transmitted per data packet delivered: ODMRP transmits more packets on redundant paths than ADMR. Therefore, as shown in Fig. 11.3d, ADMR needs to send less number of packets as compared to ODMRP. At the same time, it can be observed that the total number of packets transmitted does not vary significantly due to change in mobility. That is, the mobility does not contribute much to number of packets transmission in either of the protocols.

Forwarding efficiency: As given in Fig. 11.3e, the number of forwards is more in case of ODMRP as compared to ADMR due to excessive redundancy maintained in ODMRP. When the mobility increases, the number of intermediate transmissions increases.

Delivery latency: The delivery latency is more in case of ADMR as there are less number of redundant paths as compared to ADMR. Also, it is evident from Fig. 11.3f that the variation is more due to mobility in case of ADMR as compared to ODMRP.

11.5.2 Number of Senders

In this set of experiments, the number of senders is varied and the above parameters are measured. The multicast group size is kept fixed at 20. Node mobility is kept as low as 1 m/s. The network traffic load is 4 packets/s. The number of multicast senders is varied from 1 to 20. A single sender represents a class room scenario, whereas 20 senders represents a conferencing scenario.

Packet delivery ratio: Packet delivery ratio is studied as a function of the number of multicast senders. As the number of sources increases, performance of both the protocols degrades as depicted from Fig. 11.4a. The performance of ODMRP is almost similar to ADMR.

Number of data transmissions per data delivery to destinations: The ODMRP transmits much more data than delivered. It increases almost exponentially as

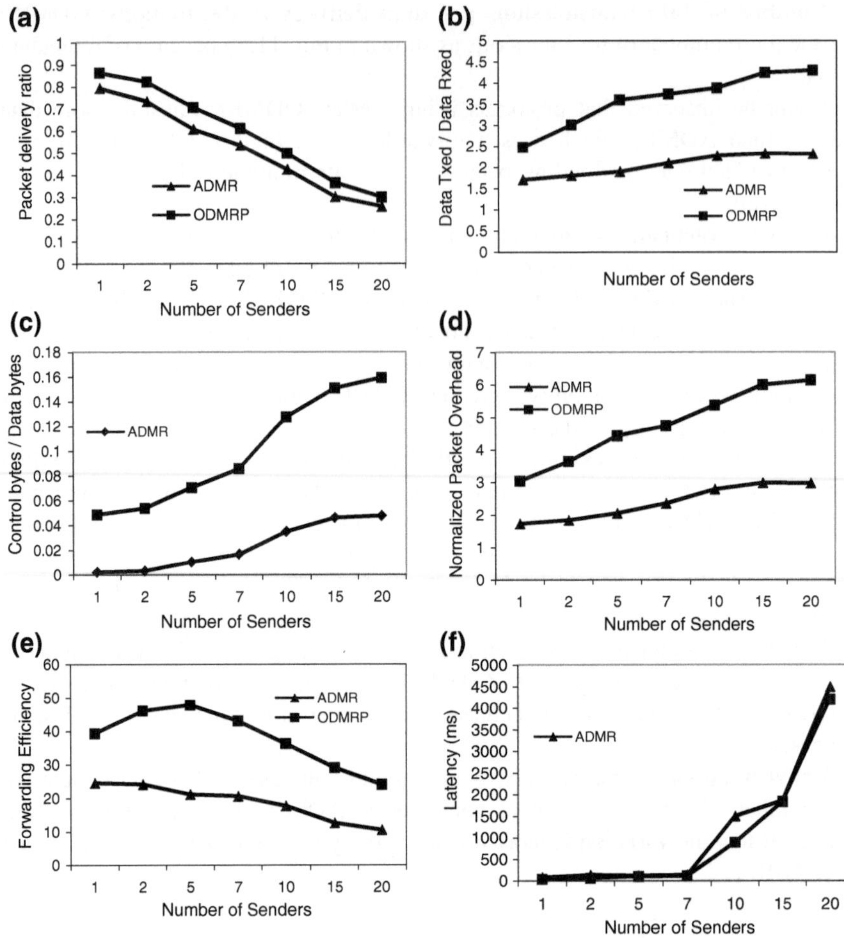

Fig. 11.4 Performance of protocols with changing number of multicast senders per group

the number of sender increases as shown in Fig. 11.4b. ADMR remains almost constant throughout.

Control overhead per data byte delivered: ADMR protocol shows a nearly constant value, while the ODMRP builds per-source meshes. If the number of senders increases, more JOIN DATA packets are propagated and control overhead grows accordingly. We can conclude from the result given in Fig. 11.4c that ODMRP in its present form may not be as efficient in networks where a large number of nodes (e.g., hundreds and thousands) are multicast sources.

Total number of packets transmitted per data packet delivered: As in Fig. 11.4d, ODMRP needs to send redundant control packets to maintain the mesh; thus, total number of packets is higher than ADMR.

Forwarding efficiency: As shown in Fig. 11.4e, the forwarding efficiency of both the protocol drops as the number of senders increase. But in case of ODMRP, it is still high due to redundancy maintained in the protocol. In case of ADMR, the efficiency degrades more as compared to ODMRP.

Delivery latency: Again, the delivery latency is shown as almost equal in both the cases in Fig. 11.4f. But ADMR incurs more latency than ODMRP as the number of senders increases. In case of conferencing kind of applications, when all needs to send simultaneously, then the delivery latency suddenly rises to very high value. Thus, none of the protocols discussed in this investigation is found suitable for this kind of application.

11.5.3 Multicast Group Size

In this set of experiments, the number of receivers is varied and the respective performances are studied. The multicast group members are varied from 1 through 20 to investigate the scalability of the protocol. Node mobility is kept as low as 1 m/s. The network traffic load is 4 packets/s. The number of multicast senders is kept 1.

Packet delivery ratio: As the number of receivers increases, performance of both the protocols degrades slightly and than remains constant as shown in Fig. 11.5a. The performance of both the protocols is almost similar and is not dependent on number of receivers.

Number of data transmissions per data delivery to destinations: The ODMRP transmits some more data than delivered as compared to ADMR. It becomes balanced for both when the number of receivers increases as shown in Fig. 11.5b. As one transmission delivers to multiple destinations, the number of data transmissions becomes almost constant with increase in number of receivers.

Control overhead per data byte delivered: Both the protocols show a nearly constant value in Fig. 11.5c. That is, it is independent of number of receivers. But due to redundancy, the control bytes are more in case of ODMRP than ADMR.

Total number of packets transmitted per data packet delivered: As in Fig. 11.5d, the ODMRP needs to send redundant control packets to maintain the mesh; thus, total number of packets is more than ADMR. But in both cases, the values are normalized when the number of receivers increases, as single transmission delivers to multiple destinations.

Forwarding efficiency: As shown in Fig. 11.5e, the forwarding efficiency of both the protocol drops increases as the number of receivers increase. But in case of ODMRP, it is more than ADMR due to redundancy maintained in the protocol.

Delivery latency: Again, the delivery latency shown in Fig. 11.5f is much more in case of ADMR. The latency in both increases as the number of receiver increases. In ADMR, latency is exponential for large number of receivers as it does not uses different paths rather it uses the fixed tree branches.

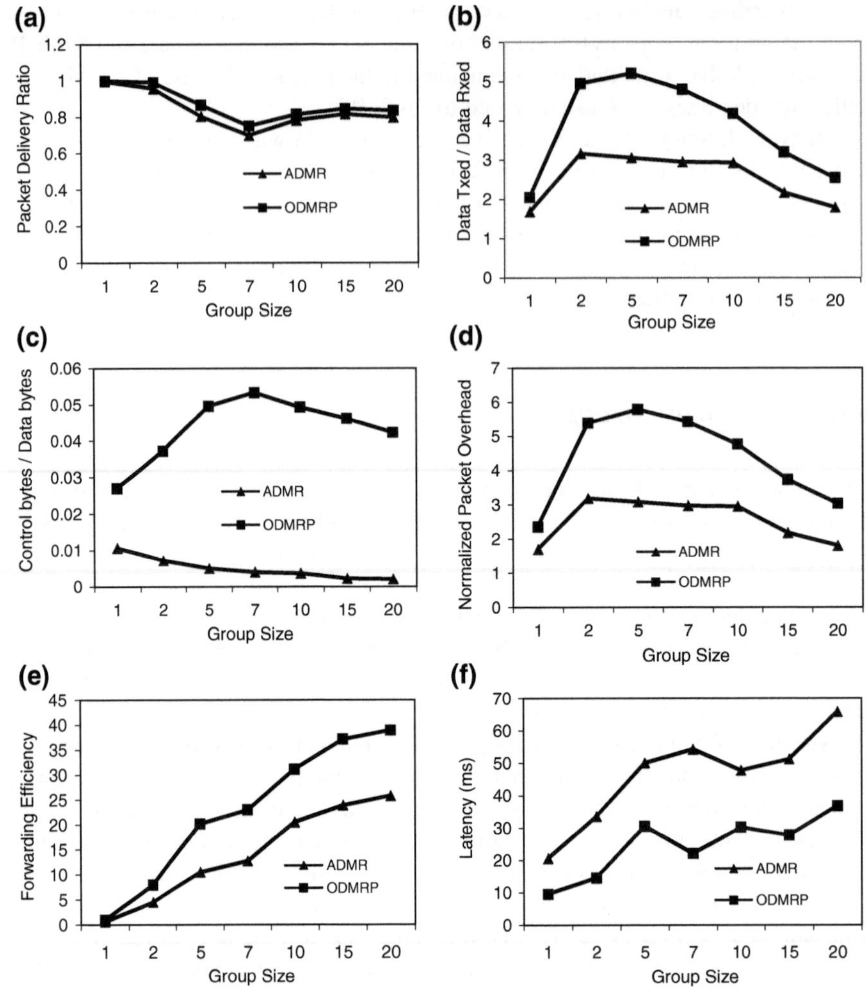

Fig. 11.5 Performance of protocols with changing number of multicast receivers per group

11.5.4 Network Traffic Load

To investigate the impact of data traffic on multicast protocol, the packet rate is varied. The multicast group size is kept fixed at 20 with one sender. Node mobility is kept as fixed at 1 m/s. The network traffic load is varied from 1 to 36 packets/s. The number of receiver is also kept fixed at 20.

The packet delivery ratio: The packet delivery load is presented in Fig. 11.6a for various traffic loads. Both the protocols are affected by load. The performance is almost similar in both the cases.

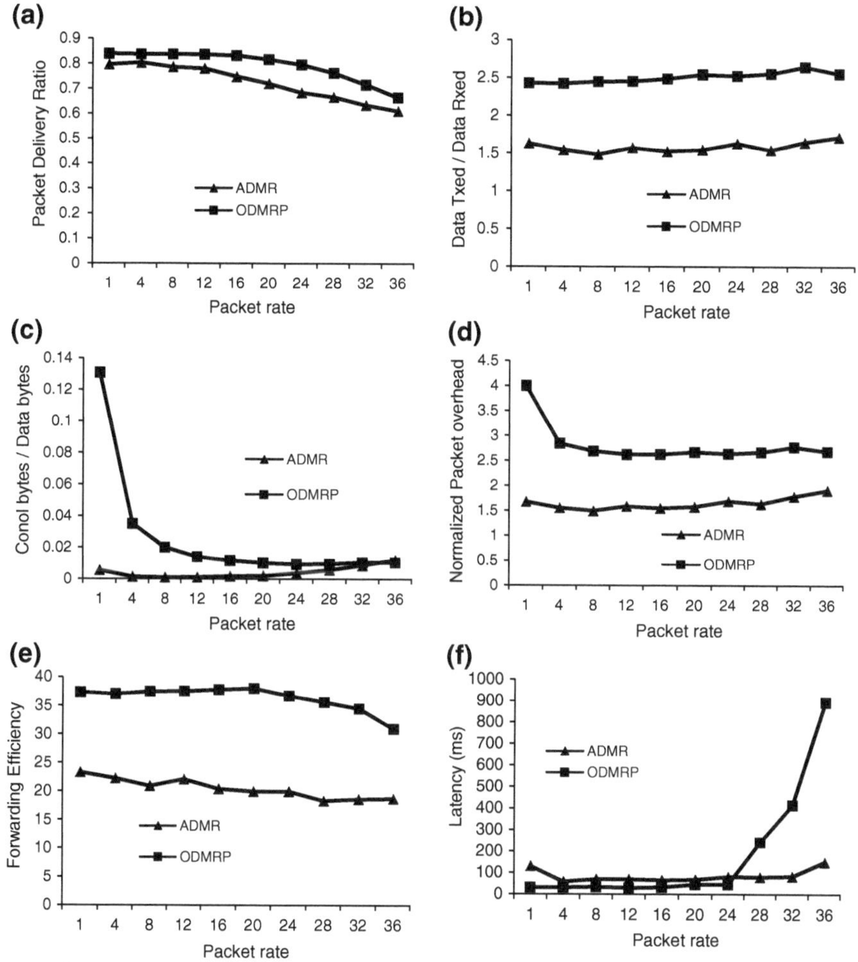

Fig. 11.6 Performance of protocols with changing network load

Number of data transmissions per data delivery to destinations: The ODMRP transmits some more data than delivered as compared to ADMR. Both the protocols show independent behavior as compared to network traffic as shown in Fig. 11.6b. That is, both values remains constant irrespective of data traffic load.

Control overhead per data byte delivered: ODMRP shows less control packet as the data traffic increases. It may be due to the fact that when there is sufficient amount of data packet, there is no need of control packets for periodic updation. In Fig. 11.6c. it is clear that ODMRP performs better than ADMR, whereas ADMR maintains a constant value irrespective of network load.

Total number of packets transmitted per data packet delivered: As shown in Fig. 11.6d, both the protocol maintains constant value regardless of data traffic. However, the ODMRP needs to send redundant control packets to maintain the mesh; thus, total number of packets is more than ADMR. But in both cases, the values are normalized when the number of receivers increases, as single transmission delivers to multiple destinations.

Forwarding efficiency: As shown in Fig. 11.6e, the forwarding efficiency of both the protocol drops as the data traffic increase. But in case of ODMRP, it is more than ADMR due to redundancy maintained in the protocol.

Delivery latency: The delivery latency is shown in Fig. 11.6f. In case of ODMRP, the latency increases abnormally as the load increases to 25 packets/s. In case of ADMR, the latency remains almost constant. It can be concluded that ODMRP is not suitable for high data traffic.

11.5.5 Pause Time

Pause time is the time a node remains stationary, after which it starts moving again. If a node moves from an arbitrary position 'a' to position 'b' and again to position 'c', then a pause time 100 means the node waits at position 'a' for a period of 100 s before moving to position 'b', similarly the node remains stationary for 100 s at node b before moving towards position 'c' again. To investigate the impact of mobility on multicast protocol, the pause time is varied from 0 to 900. A pause time 0 represents completely mobile node, whereas pause time 900 represents a completely stationary node.

In our experiment, three groups are used with 3 sources and 20 receivers per group. Node mobility is kept fixed at 20 m/s. The network traffic load is 4 packets/s.

The packet delivery ratio: The packet delivery load is demonstrated in Fig. 11.7a for various pause intervals. Though ODMRP shows better result in complete mobility mode, the performance gradually decreases as node becomes stationary. It is because of congestion caused by redundant traffic. ADMR maintains a constant value irrespective of node mobility. Thus, ADMR is more stable in all the cases considered.

Number of data transmissions per data delivery to destinations: The ODMRP transmits some more data than delivered as compared to ADMR. Both the protocols show a decreasing value as shown in Fig. 11.7b. This is due to the fact that, while node becomes less mobile, it needs less transmissions for both the protocols.

Control overhead per data byte delivered: The number of control bytes as compared to data bytes is much more in ODMRP as shown in Fig. 11.7c. Gradually, it decreases as the node becomes less mobile. In case of ADMR, these values are negligible.

Total number of packets transmitted per data packet delivered: As shown in Fig. 11.7d, the ODMRP needs to send redundant control packets to maintain the mesh; thus, total number of packets is more than ADMR. But in both cases, the values are reduced when the node becomes stationary.

Forwarding efficiency: As shown in Fig. 11.7e, the forwarding efficiency of both the protocol drops as pause time increases. But in case of ODMRP, it is more than ADMR due to more number of intermediate forwards than ADMR.

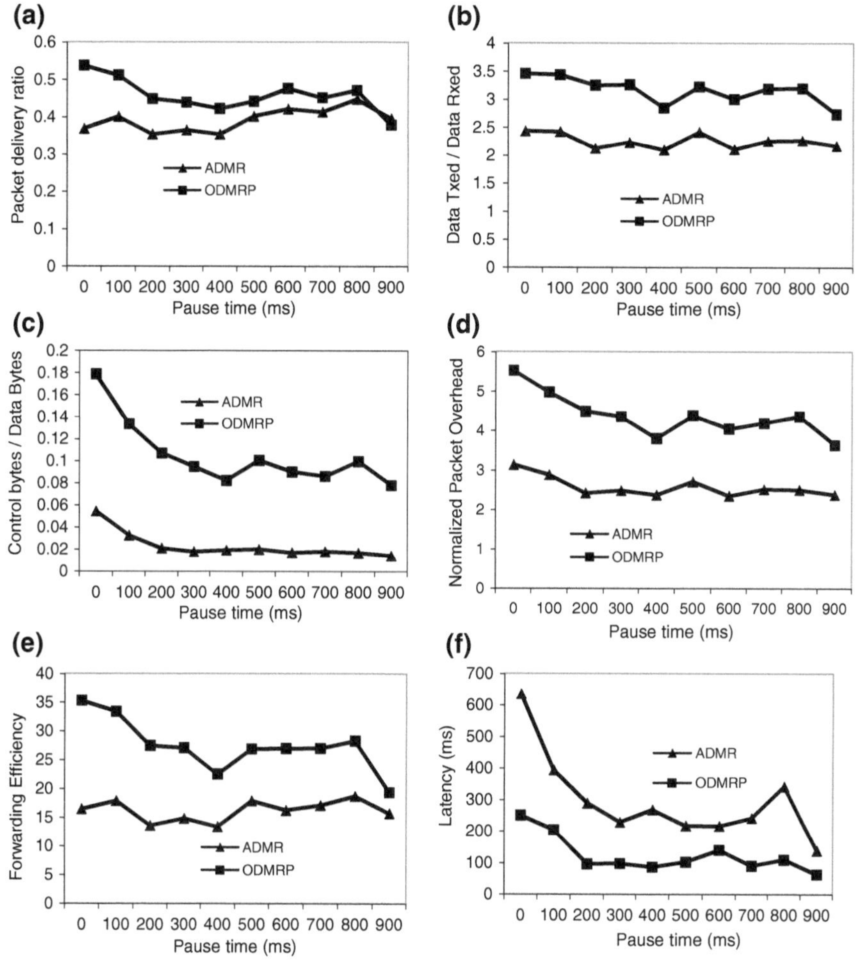

Fig. 11.7 Performance of protocols with changing pause time

Delivery latency: The delivery latency is shown in Fig. 11.7f. In both cases, latency decreases as the pause time increases. In case of ADMR, the initial latency is very high then drops suddenly when node becomes stationary for some time.

11.5.6 Number of Groups

Final set of experiments are done by varying the number of groups. In this scenario, the simulation is performed for various numbers of groups starting from one to seven. In each case, the number of sources per group is 7 and number of receivers per group is 10. The pause time is fixed at 0. Nodes move in a constant speed of 20 m/s. The data traffic is 4 packets/s.

The packet delivery ratio is shown in Fig. 11.8a for different number of groups. Both the values decrease as the number of groups increases. In case of larger number of groups, the participation of nodes in the communication increases. In case of 7 numbers of groups, almost all the nodes are engaged in communication; thus, congestion becomes high, and hence, the packet drop increases. This results in less packet delivery ratio.

Number of data transmissions per data delivery to destinations: The OD-MRP transmits 55 % more packets than delivered, whereas in ADMR, it is 30 %

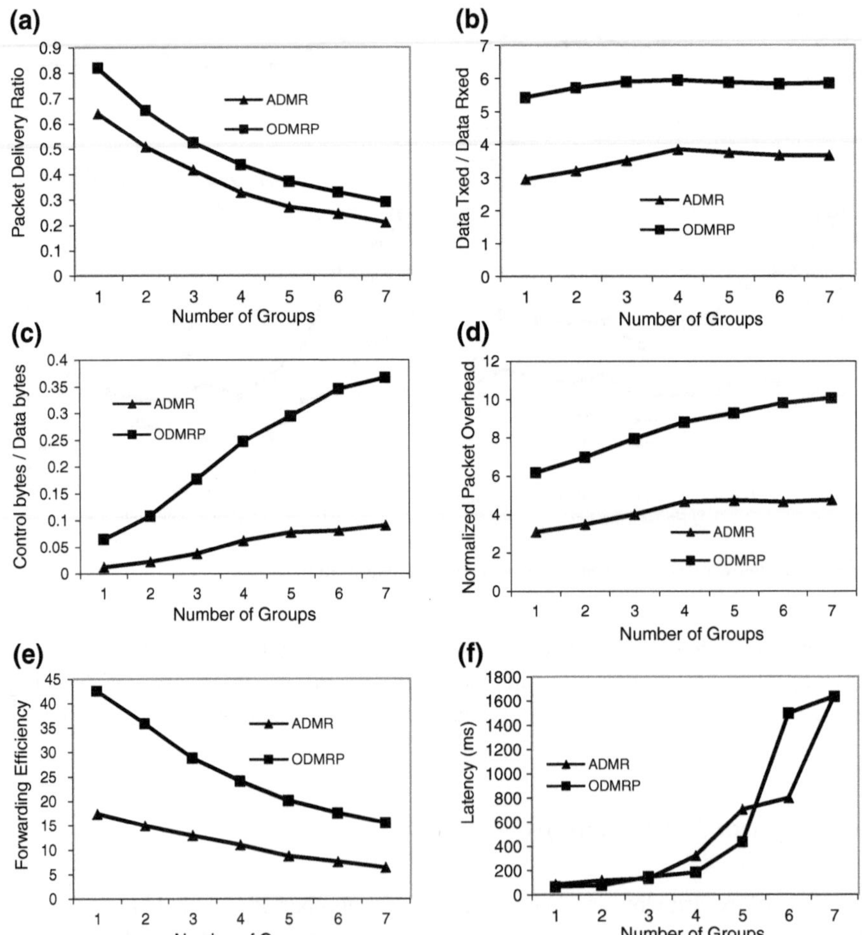

Fig. 11.8 Performance of protocols with changing number of multicast groups

data than delivered initially. But it increases further when number of groups is increases as shown in Fig. 11.8b. ODMRP imposes heavy load in the network.

Control overhead per data byte delivered: The number of control increases rapidly for ODMRP. It becomes as high as 35 % more than the data bytes delivered as shown in Fig. 11.8c. In case of ADMR, it is quite low and not varies vigorously as the participation increases.

Total number of packets transmitted per data packet delivered: As shown in Fig. 11.8d, the ODMRP needs to send more than 3 times of packets as compared to ADMR initially. Again, the total packets increase up to 10 times of the packets delivered.

Forwarding efficiency: As shown in Fig. 11.8e, the forwarding efficiency of ODMRP drops rapidly as the number of groups increases. But in case of ADMRP, the drop is smooth and it has very less number of forwards in case of more number of group

Delivery latency: The delivery latency for both the protocols is given in Fig. 11.8f. The performance is almost similar for both the protocols. When the number of groups increased beyond a certain limit (5 in this case), the performance degrades drastically and they need abnormally more time to deliver. That is, in case of more number of multicast groups/users, both the protocols are unable to perform due to heavy congestion in the network.

11.6 Performance Comparisons

The performance evaluation of two well-known multicast protocols ADMR and ODMRP has been carefully evaluated.

Figures 11.3 through 11.8 present the performance of the two protocols. Table 11.4 through Table 11.8 summarize all these results by taking average of all these values for the different considered scenarios.

As given in Table 11.4, the packet delivery ratio is 11 % less in ADMR as compared to ODMRP in case of varying mobility speed. ADMR does not perform well under high speed of the nodes due to frequent link failure and repair process. Also, poor performance may be noticed in case of varying number of groups. Thus, performance of ADMR is not as good as ODMRP under high-load and high-mobility situations. But performance of ADMR is less affected as compared to ODMRP in case of varying number of receivers per group.

However, ODMRP uses 1.6 times more forwarding than ADMR as given in Table 11.5 to obtain this high packet delivery ratio. Again, ODMRP incurs close to twice as much overhead as ADMR (Table 11.7) due to frequent flooding in the network.

The overall control bytes needed 9 % of data packet in case of ODMRP (Table 11.6), but it is as low as 2 % in case of ADMR which is almost 5 times less than ODMRP. This increase in control bytes in case of ODMRP is due to its operational need as it has to flood control packets to recreate the multicast

Table 11.4 Average packet delivery ratio as percentage of originated packets

	Range		Protocols		
Scenarios	Min	Max	ODMRP (%)	ADMR (%)	Difference (%)
Varying mobility speed	1 m/s	35 m/s	83	72	11
Varying number of senders/group	1	20	59	52	7
Varying number of receivers/group	1	20	87	84	3
Varying packet rate	1 packets/s	36 packets/s	79	72	7
Varying pause time	0 s	900 s	46	39	7
Varying number of groups	1	7	49	38	11
Overall	–	–	67	60	7

Table 11.5 Average data packets transmitted/data packets received

	Range		Protocols		
Scenarios	Min	Max	ODMRP	ADMR	Ratio
Varying mobility speed	1 m/s	35 m/s	2.6	1.8	1.4
Varying number of senders/group	1	20	3.6	2.1	1.7
Varying number of receivers/group	1	20	3.8	2.5	1.5
Varying packet rate	1 packets/s	36 packets/s	2.5	1.6	1.6
Varying pause time	0 s	900 s	3.2	2.3	1.4
Varying number of groups	1	7	5.8	3.5	1.6
Overall	–	–	3.6	2.3	1.6

forwarding state. This mechanism helps to recover from link breaks, but at the same time, it increases the control overhead and hence the battery energy consumption in intermediate nodes. In case of ADMR, the small percentage of control overhead arises due to the tree repair process in case of link breaks. The difference between the two mechanisms is that ODMRP needs control packets irrespective of link breaks, whereas ADMR needs to flood control packets as and when there is a demand for it, i.e., at the time of link brakes only.

Due to rebuilding its forwarding tree, ODMRP needs a large amount of overhead to redundant data packet forwarding (Table 11.7); whereas ADMR forwards each data packet 17 times on an average in these scenarios, ODMRP forwards each packet roughly 32 times. This huge amount of forwarding will cause unnecessary battery energy loss in the intermediate nodes.

Table 11.6 Average transmitted control bytes/data bytes

	Range		Protocols		
Scenarios	Min	Max	ODMRP (%)	ADMR (%)	Ratio
Varying mobility speed	1 m/s	35 m/s	4	0.5	8
Varying number of senders/group	1	20	10	2.3	4.3
Varying number of receivers/group	1	20	4.4	0.5	8.7
Varying packet rate	1 packets/s	36 packets/s	2.6	0.4	6.2
Varying pause time	0 s	900 s	10.5	2.3	4.5
Varying number of groups	1	7	23	6	4
Overall	–	–	9	1.9	4.7

Table 11.7 Average forwarding efficiency

	Range		Protocols		
Scenarios	Min	max	ODMRP	ADMR	Ratio
Varying mobility speed	1 m/s	35 m/s	38.8	28.7	1.6
Varying number of senders/group	1	20	38	18.8	2
Varying number of receivers/group	1	20	22.8	14.1	1.6
Varying packet rate	1 packets/s	36 packets/s	36.3	20.4	1.8
Varying pause time	0 s	900 s	27.5	16.2	1.7
Varying number of groups	1	7	26.3	11.3	2.3
Overall	–	–	31.6	17.4	1.8

Also, ODMRP creates forwarding state within nodes in the network that expires after a fixed time-out. This mechanism is responsible for the creation of redundancies in the network, since some new nodes may become forwarders for a group in current period, while forwarders created during a previous periodic flood still continue to maintain their forwarding state and thus overhear packets for that group.

Unlike the designers of ADMR [2], a different result has been discovered in case of delivery latency, though the overall performance of two protocols is same but it varies drastically at times. As can be seen from Table 11.8, ADMR is highly sensitive to mobility. It takes roughly twice the time than ODMRP to deliver the packet. Either it is a variation of mobility speed of individual nodes or pause time becomes less, ADMR is unable to perform because of frequent link breaks due to mobility and followed by a repair process, whereas ODMRP achieves network stability using redundant paths.

Table 11.8 Average latency

	Range		Protocols		
Scenarios	Min	Max	ODMRP (1)	ADMR (2)	Ratio (2/1)
Varying mobility speed	1 m/s	35 m/s	44.9	101.7	2.3
Varying number of senders/group	1	20	1036.8	1229.1	1.1
Varying number of receivers/group	1	20	24.5	46.2	1.9
Varying packet rate	1 packets/s	36 packets/s	180	85	0.5
Varying pause time	0 s	900 s	308.2	6487.7	2.1
Varying number of groups	1	7	578	543.2	0.9
Overall	–	–	435.8	442	1.0

In case of varying packet rates, ADMR performs better than ODMRP. The ADMR in this case obtains stability in its performance, whereas after a certain point, ODMRP protocol incurs a heavy delay (Fig.11.6f). When the packet increases beyond 25 packets/s, the network becomes heavily loaded in case of ODMRP due to the redundant floods/forwards of the packets. The intermediate nodes became congested and were unable to forward the packets further. Though congestion in ADMR network plays an important role in forwarding, manageable as flooding is done in the tree branches only.

A general conclusion is that in a mobile scenario, mesh-based protocols perform better than tree-based protocols in terms of packet delivery ratio, forwarding efficiency, and delivery latency. A possible reason behind this is that the availability of alternate routes provided robustness to nobility. However, ODMRP as compared to ADMR showed a trend of rapidly increasing overhead as the number of senders increases, and also, the latency increases rapidly as the load increases. ODMRP generates very large number of control packets that often lead to congestion in the network. Therefore, this protocol is not suitable for networks involving large number of groups having heavy data traffic. ADMR is good for high-load situations as it generates less control overhead, but the packet delivery ratio is comparatively low. This is due to the fact that non-availability of the route at times. But it seems ADMR is more scalable than ODMRP.

The results obtained by us are more or less similar to that of [2]. However, our results disagree those in [2] with respect to latency in case of high mobility. Our experiments show that ODMRP consumes much less latency as compared to ADMR when the nodes are highly mobile (i.e., pause time is less), or the mobility speed of the nodes is high. But ADMR takes less time than ODMRP in case of high packet rates (Table 11.8).

11.7 Summary

None of these protocols discussed here considers the load in the network while forwarding packets. That is, neither of the protocols considers the load in the intermediate nodes while the data packet is forwarded. ODMRP chooses arbitrary paths from the forwarding mesh, whereas ADMR chooses the shortest tree branch. It may happen that the shortest branch may be overloaded, while the other paths remain under loaded. This may cause a load imbalance in the network that may result in more battery power consumption in overloaded path and reduced performance. If the load can be balanced in the network, that is, each node in the network may experience almost same load at any given time, then loss of battery energy will also be uniform across the network and congestion in the network will be much less. Otherwise, as we have seen from the discussions in above sections, some of the nodes may exhaust the battery energy, and hence, they will be out of the network very soon. Also, due to load imbalance, some of the nodes may be heavily congested and become unable to forward packets and hence a reduced performance. This motivates us to design a multicast protocol to balance the load in the network as well as the control overhead while not compromising with the performance.

References

1. Lee, S.J., Gerla, M., Chiang, C.C.: On demand multicast routing protocol. In: Proceedings of The IEEE Wireless Communications and Networking Conference (WCNC), pp. 1298–1302. LA (1999)
2. Jetcheva, J.G., Johnson, D.B.: Internet draft "The adaptive demand-driven multicast routing protocol for mobile ad hoc networks (ADMR)". http://tools.ietf.org/html/draft-ietf-manet-admr-00
3. Lee, S.J., Su, W., Hsu, J., Gerla, M., Bagrodia, R: A performance comparison study of ad hoc wireless multicast protocols. In: Proceedings of IEEE Conference on Computer Communications (INFOCOM), pp. 565–574. Tel Aviv, Israel (2000)
4. Broch, J., Maltz, D.A., Johnson, D.B., Hu, Y.C., Jetcheva, J.G.: A performance comparison of multi-hop wireless ad hoc network routing protocols. In: Proceedings of the Fourth Annual ACM/IEEE International Conference on Mobile Computing and Networking, pp. 85–97 (1998)
5. Bunchua, S., Toh, C.-K.: Performance evaluation of flooding-based and associativity-based ad hoc mobile multicast routing protocols. In: Proceedings of IEEE Wireless Communications and Networking Conference (WCNC) (2000)
6. Gold Beulah Patturose, J., Immanuel Vinoth, P.: Performance analysis of multicast routing protocols IMAODV, MAODV, ODMRP and ADMR for MANET. Int. J. Adv. Res. Comput. Eng. Technol. (IJARCET), **2**(3), 1122–1127 (2013)
7. Park, V.D., Corson, M.S.: A highly adaptive distributed routing algorithm for mobile wireless networks. In: Proceedings of the IEEE Conference on Computer Communications (INFOCOM), pp. 1405–1413. Kobe, Japan (1997)
8. Pei, G., Gerla, M., Hong, X.: LANMAR: Landmark routing for large scale wireless ad hoc networks with group mobility. In: Proceedings of the ACM/IEEE Workshop on Mobile Ad Hoc Networking and Computing (MOBIHOC), pp. 11–18. Boston (2000)

9. Pei, G., Gerla, M., Hong, X., Chiang, C.-C.: A wireless hierarchical routing protocol with group mobility. In: Proceedings of the IEEE Wireless Communications and Networking Conference (WCNC), pp. 1538–1542. New Orleans (1999)
10. Raju, J., Garcia-Luna-Aceves, J.J.: A new approach to on-demand loop—free multipath routing. In: Proceedings of the IEEE International Conference on Computer Communications and Netwoeks (ICCCN), pp. 522–527. Boston (1999)
11. Chiang, C.-C., Gerla, M.: On-demand multicast in mobile wireless networks. In: Proceedings of the IEEE International Conference on Network Protocols (ICNP), pp. 262–270. Austin (1998)
12. Coreson, M.S., Batsell, S.G.: A reservation-based multicast (RBM) routing protocol for mobile networks: initial route construction phase. ACM/Baltzer Wireless Netw. **1**(4), 427–450 (1995)
13. Garcia-Luna-Aceves, J.J., Madruga, E.L.: The core-assisted mesh protocol. IEEE J. Sel.Areas Commun. Wireless Ad Hoc Netw. **17**(8), 1380–1394 (1999)
14. Wu, C.W., Tay, Y.C.: AMRIS: A multicast protocol for ad hoc wireless networks. In: Proceedings of the IEEE Military Communications Conference (MILCOM), pp. 25–29. Atlantic City (1999)
15. Gupta, S.K.S., Srimani, P.K.: An adaptive protocol for reliable multicast in mobile multi-hop radio networks. In: Proceedings of the IRRR Workshop on Mobile Computing Systems and Applications (WMCSA), pp. 111–122. New Orleans (1998)
16. Ji, I., Corson, M.S.: A lightweight adaptive multicast algorithm. In: Proceedings of the IEEE Global Telecommunications Conference (GLOBECOM), pp. 1036–1042. Sydney (1998)
17. Lee, S., Kim, C.: Neighbor supporting ad hoc multicast routing protocol. In: Proceedings of the ACM/IEEE Workshop on Mobile Ad Hoc Networking and Computing (MOBIHOC), pp. 37–44. Boston (2000)
18. Lin, C.R., Chao, S.-W.: A multicast routing protocol for multihop wireless networks. In: Proceedings of the IEEE Global Telecommunications Conference (GLOBECOM), pp. 235–239. Rio de Janeiro (1999)
19. Zhou, H., Singh, S.: Content based multicast (CBM), in ad hoc networks. In: Proceedings of the ACM/IEEE Workshop on Mobile Ad Hoc Networking and Computing (MOBIHOC), pp. 51–60. Boston (2000)
20. Ozaki, T., Kim, J.B., Suda, T.: Bandwidth-efficient multicast routing protocol for ad-hoc networks. In: Proceedings of the IEEE International Conference on Computer Communications and Networks (ICCCN), pp. 10–17. Boston (1999)
21. Royer, E.M., Perkins, C.E.: Multicast operation of the ad-hoc on-demand distance vector routing protocol. In: Proceedings of the ACM/IEEE International Conference of Mobile Computing and Networking (MOBICOM), pp. 207–218. Seattle (1999)
22. Sinha, P., Sivakumar, R., Bharghavan, V.: MCEDAR: multicast core-extraction distributed ad hoc routing algorithm. In: Proceedings of the IEEE Wireless Communications and Networking Conference (WCNC), pp. 1313–1317. New Orleans (1999)
23. Wieselthier, J.E., Nguyen, G.D., Ephremides, A.: Algorithms for energy-efficient multicasting in ad hoc wireless networks. In: Proceedings of the IEEE Military Communications Conference (MILCOM), pp. 1414–1418. Atlantic City (1999)
24. Acharya, A., Badrinath, B.R.: A framework for delivering multicast messages in networks with mobile hosts. ACM/Baltzer Mobile Netw. Appl. **1**(2), 199–219 (1996)
25. Bartoli, A.: Group based multicast and dynamic membership in wireless networks with incomplete spatial coverage. ACM/Baltzer Mobile Netw. Appl. **3**(2), 175–188 (1998)
26. Royer, E., Perkins, C.E.: Multicast operations of the ad hoc on demand distance vector routing protocol. In: Proceedings of the 5th ACM/IEEE Annual Conference on Mobile Computing and Networking, pp. 207–218 (1999)
27. Ji, L., Corson, M.S.: Differential destination multicast (DDM) specification. Internet-Draft, draft-ietf-manet-ddm-00.txt, July (2000) (Work in progress)
28. Vaishampayan, R., Garcia-Luna-Aceves, J.J., Obraczka, K.: An adaptive redundancy protocol for mesh based multicasting. Elsevier Sci. Comput. Commun. **30**, 1015–1028 (2006) (online)

29. Kaplan, E.D. (ed.): Understanding the GPS: Principles and Applications. Artech House, Boston (1996)
30. Ballardie, T., Francis, P., Croweroft, J.: Core based trees (CBT)-An architecture for scalable inter-domains multi test routing. In: Proceedings of the ACM SIGCOMM Symposium on Communications Architectures, Protocols and Applications, pp. 85–95. San Francisco (1993)
31. Bommaiah, E., Liu, M., McAuley, A., Talpade, R.: AMRoute: Adhoc multicast routing protocol. Internet-Draft, draft-talpade-manet-amroute-00.txt, August (1998) (work in progress)
32. Sandoval G, E.I., Galvan T, C.E., Galvan-Tejada, J.I.: Multicast routing and interoperability between wired and wireless ad hoc network. Procedia Eng. **35**, 109–117 (2012)
33. The network simulator—NS-2, http://www.isi.edu/nsnam/ns/
34. Wireless Multicast Extensions for ns-2. http://www.monarch.cs.rice.edu/multicast_extensions.html
35. IEEE Computer Society LAN MAN Standards Committee: Wireless lan medium access control (MAC) and physical layer (PHY) specifications, IEEE Std 802.11-1997. The Institute of Electrical and Electronics Engineers, New York (1997)

Author Index

S. Patnaik et al. (eds.), *Recent Development in Wireless Sensor and Ad-hoc Networks*,
Signals and Communication Technology, DOI 10.1007/978-81-322-2129-6